THE COMPLEXITY OF CREATIVITY

SYNTHESE LIBRARY

STUDIES IN EPISTEMOLOGY,

LOGIC, METHODOLOGY, AND PHILOSOPHY OF SCIENCE

VOLUME 258

THE COMPLEXITY OF CREATIVITY

Edited by

ÅKE E. ANDERSSON

Institute for Futures Studies,
Stockholm, Sweden

and

NILS-ERIC SAHLIN

Department of Philosophy,
Gothenburg University
and
Lund University,
Sweden

KLUWER ACADEMIC PUBLISHERS

DORDRECHT / BOSTON / LONDON

A C.I.P. Catalogue record for this book is available from the Library of Congress.

ISBN 0-7923-4346-8

Published by Kluwer Academic Publishers,
P.O. Box 17, 3300 AA Dordrecht, The Netherlands.

Kluwer Academic Publishers incorporates
the publishing programmes of
D. Reidel, Martinus Nijhoff, Dr W. Junk and MTP Press.

Sold and distributed in the U.S.A. and Canada
by Kluwer Academic Publishers,
101 Philip Drive, Norwell, MA 02061, U.S.A.

In all other countries, sold and distributed
by Kluwer Academic Publishers Group,
P.O. Box 322, 3300 AH Dordrecht, The Netherlands.

Printed on acid-free paper

Printed in the Netherlands

Table of Contents

EDITORS' PREAMBLE

This is a book on the concepts, theories, models and social consequences of creativity. The articles are the outcome of a workshop on creativity held in Venice in October 1994. The workshop was organized by the Swedish Institute for Futures Studies as part of the larger project The Cognitive Revolution.

The role of textbooks in the teaching of creativity was discussed at the workshop. One of the participants, Gudmund Smith, asserted that not reading textbooks might be the most creativity enhancing technology in education.

However, this is not a textbook, nor is it a set of simple instructions of how to become creative, but we hope that this volume contains enough of thought provoking material to spark off a creative process.

Å.E. A.
N.-E. S.

Stockholm, June 1996

MARGARET A. BODEN

THE CONSTRAINTS OF KNOWLEDGE

People sometimes speak as though knowledge has little to do with creativity, or even prevents it. This dismissive attitude toward the role of cognition in creativity takes two main forms.

On the one hand, cognition is contrasted with motivation. Naturally enough, the creative role of motivation is stressed by psychodynamicists in general. But even cognitive psychologists allow that the reasons why one individual achieves creative greatness (what I have called H-creativity (Boden, 1990, ch. 3)) while another, inhabiting the same cognitive niche, does not, lie primarily in differences in motivation (Perkins, 1981).

The precise motives may differ, but their intensity does not. H-creators are driven, and in turn drive the people in their social ambit even unto death. Florence Nightingale is a famous nineteenth-century example: lying on her sickbed, she dictated (in both senses) to her male helpers—some of whom sickened and died under the strain. Seven twentieth-century examples—Gandhi, Freud, Martha Graham, T. S. Eliot, Stravinsky, Einstein, and Picasso – have been described by Howard Gardner (1993), whose study makes it abundantly clear that H-creative personalities are, to put it delicately, "difficult". In their singlemindedness in pursuing their creative mission, and their near-total lack of concern for other people's interests, each is a match for Nightingale.

On the other hand, it is said that knowledge is unnecessary for creativity, and/or that it can stifle it at birth. Einstein (we are often reminded) was a mere clerk, not a university scientist, when he did his pioneering work. As such, he was free from the institutionalization of his motives (no fear of challenging his boss)—and of his knowledge, too. There are many other less exalted examples of "outsiders" coming up with H-creative ideas, people who were either partially ignorant of or not professionally committed to the current orthodoxy. Thomas Kuhn reminded us how difficult it is to change one's mind, especially about systematic beliefs: science changes because old scientists die (Kuhn, 1962). Similar remarks apply also to the arts, although the deadening effect of orthodoxy is arguably less marked here, because of the (relative) lack of systematicity and the post-Romantic emphasis on novelty and individual expression for their own sake.

Such views can lead, for instance, to young children being encouraged to express their own views in their writing without being taught the discipline of writing to help them do so. While this may be healthy as a temporary

Å. E. Andersson and N.-E. Sahlin (eds.), The Complexity of Creativity, 1–4.

measure, and as a reaction against pedantic grammaticism and cultural imperialism, in the long run it does the child no favours. Creativity is founded in discipline.

I do not mean the discipline of hard work (George Bernard Shaw's "99% perspiration"), though that is indeed necessary. Rather, I mean the cognitive constraints which form and inform creative ideas, and which make them possible. Much of the "99% perspiration" is needed for becoming familiar with these constraints.

To be sure, one type of creativity is relatively unconstrained. In this ("combinational") type, the valued novelty consists in an unusual combination of familiar ideas—as in poetic imagery, or analogy (Boden, 1990, ch. 6). Even here, however, not anything goes.

The poet is, if anything, even more careful than the prose-writer to find *le mot juste*. Even Joyce did not try to explode the reader's consciousness with a merely random sequence of concepts. Moreover, those artists who deliberately use random methods in generating their artworks do so only against a background of artistic discipline. This applies in all domains, from Mozart's dice-music to Brian Eno's *oeuvre*, or from the "self-contradictions" of surrealism to the DIY-assembly novel—published not as a bound volume but as sections loose in a box, which (except for the first and last) can be read in a random order (Johnson, 1973). In general, randomness (and serendipity) can contribute to creativity, but only if it can be intelligibly related to the relevant cognitive background (Boden, 1990, ch. 9; 1995).

Likewise with analogy. One can compare anything with anything else (a raven with a writing-desk, for instance). But if an analogy is to be developed for rhetorical or problem-solving purposes, the two analogues must be matched in some detail. This may require very hard thinking, by orator or oncologist.

The other type of creativity involves the exploration and transformation of conceptual spaces (Boden, 1990; 1994, ch. 4). These spaces are styles of thinking, or sets of generative constraints. They are positive constraints, not negative ones: they express what may be thought, not what must not be thought. Tacitly, however, they do constrain in the negative sense. For if an idea is unthinkable within a particular conceptual space (as the benzene-ring was, within early organic chemistry), then it cannot be thought unless that space can be changed in some way.

Superficial tweakings and fundamental transformations of the dimensions of the space are both possible. But the latter is more rare. — Why? What is the nature of the mental "inertia" which makes it more difficult to transform a conceptual space than merely to tweak it? Or, if the transformation does take place, why is it less likely to be accepted—even in a brainstorming session, where (so the instruction has it) "anything goes"?

The latter question reminds us that the very concept of *creativity* is intrinsically evaluative. To call an idea creative is to commit onseself to the view that it is both new and interesting. To be sure, what counts as interesting will differ from domain to domain, and to some extent from person to person. Accordingly, much dispute about "creativity" focusses not so much on the novelty of the disputed idea, but rather on its value. Even in the sciences, such disputes may last for many years (Schaffer, 1994). This need for evaluation helps us to see why transformation is even more problematic than tweaking.

Transformations of conceptual spaces present two difficulties, neither of which applies in such strong degree where more superficial changes are in question. First, the novel (transformed) idea may not be readily intelligible. In other words, the newly transformed conceptual space is not easily navigable, because it is too different from the previous, familiar, one. Second, a transformation may imply so many changes in the local topography of the conceptual space that it fails to meet certain (empirical or stylistic) criteria we are not prepared to give up. We may be so loath to abandon these criteria that the novel idea is rejected immediately, being regarded as absurd.

In order to judge such matters with sensitivity, and to adapt other aspects of the conceptual space so as to cohere with the surprising transformation, *expert knowledge* is needed. For even transformational creativity is not the abandonment of all constraints. (That way, lies unintelligibility.)

Deep and detailed knowledge of the conceptual space concerned is required. It is needed in evaluating the relative importance of different constraints, in choosing which ones to drop or deform in the case of a clash, and in deciding which transformations are worth following up and which are not.

Almost anyone could unthinkingly suggest applying a transformational heuristic, such as "drop a constraint" or "consider the negative". But not anyone could actually apply the heuristic: one has to have some idea of what the current constraints are, in order to drop or negate any of them. So a competent geometer (Lobachevsky, for instance) can suggest dropping Euclid's last axiom, because expert geometers know that this is the one which introduces the problematic notion of parallelism, and this is the one which seems to be conceptually independent of the other axioms. Similarly, a competent chemist (such as Kekule, puzzling over the structure of benzene) can suggest turning a string molecule into a ring molecule, because chemists know that the behaviour of a molecule depends not only on its constitutent atoms but also on their neighbour-relations, or topology. Above all, relatively few people could competently (and imaginatively) evaluate the novel result of applying the transformational heuristic.

The relation between constraints and creativity, then, is a subtle one. Constraints there must be, for it is cognitive constraints which define conceptual spaces, old and new, and which enable us to find our way through them. They function as generative structures and exploratory guidelines, without which we could neither come up with relevant new ideas nor appreciate them once we had done so. But the constraints must not be too constraining.

If they cannot be changed at all, we can have only exploratory creativity. That is not to be scorned, for many positively valued novel ideas, both in "normal" science and in relatively unadventurous art, are found by exploring some familiar conceptual space. The more complex the constraints defining that space, the more likely it is that exploration will give valued results, since the exploratory potential of the space will not have been obvious from the outset.

Perhaps the constraints can be only slightly changed, or perhaps someone can change only the relatively superficial ones—those which define near-accidental features of the space, rather than its fundamental structure. If so, then we can aspire to no more than superficial creativity, to inessential tweakings of a cognitive structure that remains fundamentally unchanged.

To generate the most surprising, and (if accepted) the most highly valued, creative ideas, one or more of the fundamental constraints must be altered so as to transform the space itself. To transform the space is not merely to make one change at a certain point in time. Rather, it is to transform the space's generative potential. New types of exploration and tweaking are now possible, though it may take many years for them to be realized.

All these forms of creativity, even "mere" exploration, require some expertise. The fact that expertise can also stifle novelty contributes to the delicate balance of creative thought. The creative person need not—perhaps should not—be a "know-all". But they cannot be a "know-nothing", either. Creativity is engendered by knowledge, in being constrained by it.

School of Cognitive and Computing Sciences,
University of Sussex, England

REFERENCES

Boden, M.A. (1990) *The Creative Mind: Myths and Mechanisms.* London: Weidenfeld & Nicolson, 1990.

Boden, M.A. (1994) "What Is Creativity?" In M.A., Boden (ed.), *Dimensions of Creativity.* Cambridge, Mass.: MIT Press. Pp. 75–118.

Boden, M.A. (1995) "Creativity and Unpredictability". In S. Franchi & G. Gulvedere (eds.), *Stanford Humanities Review.* Constructions of the Mind: Artificial Intelligence and the Humanities, vol 4, no 2, pp. 123–140.

Gardner, H. (1993) *The Creators of the Modern Era.* New York: Basic Books.

Johnson, B.S. (1973) *Aren't You Rather Young to be Writing Your Memoirs?* London: Hutchinson.

Kuhn, T.S. (1962) *The Structure of Scientific Revolutions.* Chicago: University of Chicago Press.

Perkins, D.N. (1981) *The Mind's Best Work.* Cambridge, Mass.: Harvard University Press.

Schaffer, S. (1994) "Making Up Discovery". In M. A., Boden (ed.), *Dimensions of Creativity.* Cambridge, Mass.: MIT Press. Pp. 13–52.

INGAR BRINCK

THE GIST OF CREATIVITY

Creativity is a notoriously evasive concept, and it is used to cover a lot of different phenomena. Different methods and a wide variety of angles have been used in the striving for a clear-cut conception. The focus has been on alternatively the personality of creative people, their childhood, the conditions that a society must fulfil to support a creative atmosphere, works of art contra the discoveries of science, changes in pedagogy to give rise to or improve creativity, computer models, intuition, and so on. Consequently, the resulting picture of creativity varies substantially depending on the goal of the inquiry as well as on the constraints that are set from the start, not only by the scope of the investigation, but also by the discipline that the investigator belongs to and the method that is used.

I will make use of a broad conception of creative activities. I do not think there is any difference in kind between everyday and scientific creativity or between the creativity of grown-ups and children. My examples of creativity come from cooking, architecture, science, and gardening. However, in order not to lose contact with that elusive characteristic we have in mind when we call something or somebody creative, we must pay attention to the common sense conceptions that surround creativity, and let these conceptions guide the account. The idea is to track that which lies behind all the things that get the epiteph 'creative'. According to how the word 'creative' is used in everyday language, the phenomena to which it applies should instantiate some, if not all, of the following properties: novelty[1], unexpectedness, fertility, surprise, adequacy or correctness, and finally in some sense be deliberate (as opposed to random).[2] It should also involve an active search, emanating from the efforts of the individual. An activity that can be described by these properties I will call *creative*, and I intend this use of the word to follow common usage. In a derived sense we can talk about the creative products of such activity. You will notice that I often talk about problem-solving instead of producing creative results, being creative, or similarly. The reason is that

[1] M. Boden distinguishes between person-related and historical novelty, the former being new to the person who came up with it but possibly old to other people, the latter being of historical importance. I am mainly interested in the former kind of novelty. Cf. *The Creative Mind* (1991) New York: Basic Books.

[2] Most of these properties are relational: an action or thought is unexpected, fertile, surprising, adequate, et cetera, in relation to a person or certain people in a certain field at a certain time.

Å. E. Andersson and N.-E. Sahlin (eds.), The Complexity of Creativity, 5–16.

I in fact consider creativity to be some sort of problem-solving, but of a specific kind. You will see what kind in a while.[3]

I want to start by making a distinction between three different angles on creativity: the contributions of the individual, the environment, and the knowledge domain, respectively. First, the *individual*. An inquiry that concentrates on the individual will bring up the origin or genetic causes that made that person creative. It will examine the childhood, the personality, and the personal qualities of the individual and try to determine what is required of an individual and its immediate surroundings for it to become creative. Secondly, the *environment*. The importance of the environment to creativity is manifold and can be assessed by a study of the social and cultural factors at a given point in time. For instance, a society that promotes and encourages creativity and strives for novelty and unexpected solutions to problems will probably foster more people that behave in a way that we would call creative than a society that is indifferent or negative. But what would an atmosphere that enhances creativity look like? Such a question is typical of this approach. Thirdly, we have the *knowledge domain* of the individual, characterised by a description of what the individual has knowledge of and by the kind of knowledge he or she has. Note that the general cognitive state that characterises the field in which the individual works or acts falls under the second point.

At this stage, two basic questions about the knowledge domain present themselves. One has to do with its *prevailing state*, the other with its *dynamics*. The first question is answered by facts about how well-developed and organised the domain is and about what kind of knowledge it involves. It seems that for an individual to be creative in an area, he or she must have a good grip of that area, or whatever result he or she attains, it will be random, and thus cannot be called creative. By 'having a good grip' I intend that the individual has rather detailed knowledge, whether tacit or explicit, and also has experience of working in the domain from earlier occasions. She should probably as well have some kind of background knowledge that gives a broader basis for solving the problem that she has encountered.

The second question, about the dynamics of the knowledge domain, is answered by an explanation of what has to happen in the domain if the knowledge that is represented in it shall result in creative solutions or actions. Let us say that two individuals are in the same state of knowledge with their knowledge domains organised in exactly the same way. They try to solve the same problem. Only one succeeds in giving an interesting, fruitful answer

---

[3] It should be made clear that the concept of problem solving that I use is very general and broad. Anything from childrens' play to furnishing over painting and doing laboratory work count. In all these cases, the individual stands before a problem (how to furnish the flat, how to depict the landscape, et cetera) and tries to solve it by reasoning, although the reasoning not always needs to be explicit, that is, in words.

that leads to unexpected consequences. How come? This is the question that I am interested in here. To answer that question means to reveal some of the mysteries that go by the names of intuition, association, and imagination.

First, the difference between normal problem-solving and that specific kind that we call creativity must be made clear. Both kinds involve the same components: *knowledge representations*; *rules* for manipulating those representations; a *direction* to the activity; *standards for evaluating* the results that are produced; and finally something that *puts an end* to the inquiry. It is typical for situations that require creative efforts as opposed to step-by-step reasoning that they have an open-ended character. The goal of the inquiry is not well-defined. The direction in which the answer should lie is rather evident, but the nature of the answer is unknown, and there might even not be an answer. Because of this open-ended character, the evaluation standards must be less exigent than in the normal case. They must allow for solutions that are outside the scope of the inquiry, that exceed the expected. Before a solution can be chosen, it may be necessary to match different answers with one another and see onto what they lead. Likewise, the halting rule or stop mechanism has to be more flexible in creative problem-solving than in the normal case. One should not be satisfied with any result that gives what looks like a correct answer, but wait for a result that leads the inquiry in a new direction or at least gives a new and unexpected answer.

A consequence of a creative advance is a reevaluation of old evaluation standards and halting rules. The advance not only opens up a new field, it also sets new expectations for further inquiries. This fact helps distinguish between normal, as opposed to extraordinary, problem-solving, where the problem as well as the expectations that control the inquiry are well-defined, and the steps by which the solution can be reached are well-known. Extraordinary problem-solving occurs in situations where one cannot get to the solution by using the available tools. The problem stands in need of a creative solution. It lurks on the threshold between old and sterile knowledge and new but still uncovered ground, and the search for a creative solution is accompanied by a relaxation of the given standards.

The use of special evaluation and halting rules will nevertheless be futile unless the material that they are applied on is right. And now we get to the core. I maintain that creativity results from a certain kind of operations on the knowledge representations in the domain. Unless these operations occur, whatever the psychological, social, or cultural conditions are that hold of the individual, creativity will not arise. Normally, the rules for problem-solving proceed step-by-step in the following manner. Say that an agent is in a state of knowledge S_1 and performs an action A that consists in one single operation, for instance a negation of S. Then the next state S_2 will be a function of S_1 and A. The succeeding state S_3 will in turn be a function of S_2 and a second action A_2. And so on. Of course, this is a simplified model, but I hope the main idea is clear.

Another property of rules for normal problem-solving is that they are truth-preserving. A step that does not preserve the truth of the reasoning will not be taken into account. N-E. Sahlin has suggested (personal communication) that regarding creativity other characteristics may replace truth as being more salient or valuable. Such characteristics could be pragmatic values like fertility, aestethic values like elegance, or simply dissimilarity or divergence of the content generated by the rule, or why not the property of being energizing. I will not dwell on these properties, but concentrate on the fundamental nature of the rules at work in creativity.

It seems that there are two kinds of rule that operate in creativity: *intra-representational and inter-representational.* They both result in a change in how a piece of knowledge is represented. The first kind consist in changes within a representation, while the second one involves more than one representation. As regards this latter kind, it is suitable to talk about transfers between source and target representations. What is transferred to the target is either structural traits or contents of the source, and the change pertains to the target. It so to speak 'copies' properties from the source. The transfer between source and target both rests upon and generates judgments of similarity between the domains involved.[4]

Intra-representational rules do not depend on similarity judgments. They constitute a way to reorganise or reconceive a given content. If we look closer at the intra-representational changes, there are at least three sorts. A first one consists in *changes of the form* of representations. That is, the same information can be represented in different ways, and when changes are made, the information is put in a new light that may permit of new inductions or at least a new understanding. The changes I have in mind are between symbolic (or conceptual), indexical, and iconic forms of representations.[5] It is important to notice that these changes not only concern variations in linguistic notation, but apply to all kinds of representations. It is exactly this alteration between forms that make the rules so powerful as regards creativity.

I will give a few examples. We want to signal that smoking is forbidden. We can do so by writing a sign in a natural language, or we can use a picture of a cigarette that is crossed over, or we can use some kind of very strong fan that starts every time a cigarette is lit and puts it out. The first strategy is symbolic, the second predominantly iconic, and the third one predominantly

---

[4] I distinguish between the structure and the content of representations. The structure organizes the content, while a certain content will make some structures more probable than others. Content is best described by a a list of features or characteristics. The organisation of those features is provided by a description of the structure that explains how the features are related.

[5] Indices represent by being close in time and space to the represented object and are often causally related to that object. Icons represent by having a pictorial similarity with the represented object. Symbols or concepts do not depend on either the presence or existence of what they represent. Linguistic symbols are usually considered to be conventional and arbitrary.

indexical. By changing between forms of representation in this manner one can escape from patterns of thinking that connect certain content with certain forms. Depending on how the content is represented, that is, what form and what medium are used, the same content may give rise to different thoughts and different behaviour on the part of the one who receives or entertains the information.

Say that we want to furnish a room. We can do it by simply putting the furniture in the room and moving it around till everything looks OK, or we can draw maps and make diagrams of the relation between the room and the furniture and between different pieces of furniture, or we can make drawings that depict the room with different arrangements of the furniture. We can also write lists of which pieces of furniture should be next to each other of practical or other reasons and then compare the results of the different lists. The changes can of course also pertain to different kinds of symbolic representations, as for instance Arab or Roman numerals, or linguistic or formal-logic representations of agruments and proofs.

A second sort of intra-representational rule concerns *changes of modality*, and has to do with which sensory modality carries the heaviest load in the reception of information. Say you are cooking a fish-soup. You make it with the help of a recipe. The result of the recipe is unsatisfactory, and you must improve on it. To facilitate the task, you concentrate on one aspect of the information that you have of the soup at the time. In turn, you taste it, smell it, check its consistency, and finally consider how it looks. Depending on at what your attention is directed your measures will vary.

A third sort of intra-representational rule consists in *shifts in the figure-ground organisation*. To shift your focus between the fish, the carrots, the tomatoes, the clams, and so on, of the soup will help you find new ways to improve it. Aspect-seeing can also be assigned to this group. It consists in variations of how a representation is understood depending on what aspect of it is highlighted. A famous example which most of us are familiar with was used by Wittgenstein: the duckrabbit.

As regards the inter-representational rules, I mentioned above that they involve a transfer between source and target representations. A first sort *transfers structure*. You have the basic material for solving the problem, but you do not know what to do with it. You need a way to organise the different elements of your representation. I will give three examples of this. First, Kekulé and the structure of the benzene molecule. According to the chemistry of that time, all organic molecules should be possible to describe in terms of a string of carbon atoms. But the proportion of the elements of the benzene molecule seemed to make such a description impossible. Kekulé was dozing by the fire when he woke up with a sudden insight. The form of the snake biting its tail that he saw in his mind's eye was also the form of the benzene molecule. The molecule had a ring- and not a string-structure. To proceed to explain the structure of the benzene molecule it was not sufficient to reconceive its form from being linear to circular. Additional changes in the model had to be made, some of which could not be verified at the time. All the changes were, however, governed by the model based on the snake-vision.

A similar example concerns the model of the atom which at the beginning of this century came to be conceived of as a miniature solar system: the atom as a microscopic sun and the electrons circling around it as the planets. This model was contradicted by many established theories of the time, for instance in electromagnetics, but Bohr stuck to it. Gradually he succeeded in working out a new and successful theory of the atom based on the model. These two examples show that at times when no solution is at hand, the search may fruitfully be guided by a model or structure from an area apparently unrelated to the main subject.

A final example is drawn from architecture. On innumerable occasions architecture has been governed by models taken from other areas, and whose primary features have not had much to do with the comfort of living or building of houses. As one case Le Corbusier can be used. He wanted his housing areas to be biological or ecological systems, in which nature and housing supposedly had made a pact. New buildings were seen as plants that were introduced in the existing scenery. To Le Corbusier, constructions should be part of the life cycle, not stand outside it as static, dead objects.[6] Visions like these governed the planning of new areas and the calculations of new constructions. A second case is Brazilia, the capital of Brazil, that was created from scratch in the end of the 1950s by Costa and Niemeyer. It was guided by a vision of the future of the country, and it is built to resemble a jet plane from the air (that supposedly flies into the future). Unfortunately, the futuristic design appears to have been ahead of its inhabitants in time, who were not mentally (nor socioeconomically) ready to give up the more homey conglomeration of the modern city for a visionary, clean-cut and ultrafunctional world capital.

Let us proceed to inter-representational rules that *transfer content* between source and target domains. Examples can easily be found of situations in which you know how to do, but not what to use to do it. A first one I draw from grafting. Usually when a branch is grafted onto a tree, say an apple-tree, bast is used to keep the branch in place. A friend of mine was grafting but discovered in the last minute that he was out of bast. What should he do? Usually, when you want something to stick to another thing you use either glue or adhesive tape. Glue was not practicable in this case, but tape was. My friend used masking tape, which not only held the branch in place, but turned out to have better qualities than bast. It is stretchable and it does not burst because of bad weather conditions and thus does not have to be exchanged. Besides, the grafting was hundred per cent successful, a strikingly

---

[6] See for instance his *La Maison des Hommes* (1936) Paris: Librairie Plon.

good result, and according to my friend a consequence of his using masking tape instead of bast. This is a case in which the function of the source and target domains coincide, while the qualities differ. Bast and masking tape have similar functions, at least in the case of grafting, and that led my friend to change the bast for the tape. They have, however, different properties.

A second example comes from the development of the theory of evolution. The problem concerns how to explain the creation in terms of evolution.[7] How should evolution be understood in order to give such an explanation? A proper description of what evolution involves was lacking. Both the traditions from early Greek philosophy and from Christianity conceive of the world as set. No qualitative evolution occurs, everything is already in place, created at one single point in time. In European 17th and 18th century thought, however, a new conception of change had developed. Science had made many improvements since the Middle Ages, and the belief in natural laws that govern the continuous progress of man grew strong. In the middle of the 18th century in France, the idea of progress was for the first time applied not only to man and to cosmos, but also to nature. This idea took its time to strike root, since it lead in the direction of a denial of God as the Creator. During the 19th century it slowly became accepted, but the thought about the constancy of the species still dominated. The impetus of Darwin to advocate the evolution of the species came from a work in geology. By the time Darwin was working on *On the Origin of Species* others came up with similar, but less detailed, ideas. One of the most important general ideas that were put forward in Darwin's book dealt with the design with which evolution took place. Darwin did not attribute that property to God, but introduced the principle of natural selection.

What conclusion concerning creativity can we draw from all this? No doubt, Darwin counts among the people that almost everybody would call creative. The interesting fact behind this story is that Darwin did not create anything from scratch. The idea about the evolution of the species grew slowly over the centuries, and over time borrowed its fundamental features from other areas, from comparisons of nature with man, with cosmos, and so on. This kind of slow development accounts for why often several people come up with the same idea independently of each other. It is not primarily the stimulating environment that provides the impulse for a new sort of solution to a problem, but prior comparisons between different knowledge domains, and the resulting transfer of features from source to target.

A third example of inter-representational transfer of content is the comparisons between man and machine. These go in both directions, depending

[7] If the problem was how to explain the presence of organisms on earth we would have a case of transfer of structure, since we would know what we wanted to do, but not which explanation should be used or from what area it should be taken. To solve the problem we could for instance choose between creationism or Darwinism.

on the subject of interest. When we do research on intelligence and reasoning, we take the way the computer works as a model. When we search for new solutions within robotics, the way perception and motor activity function in man constitutes the model. In the first case, we need something new to fill in the model of what it is to think, in the second case we need an account of sensory and motor control of robots that does not take its starting-point in our knowledge of machines. A related example concerns the study of animals either from the view-point of animals as machines in the tradition of Descartes or as beings with a soul—the latter view verging on anthropomorphism.

The last examples all depict how characteristics pertaining to one domain are transferred to another. I have not discussed cases in which several domains are involved, since that would unnecessarily complicate the survey. I am sure, however, that such cases can be found. Above I used as wide a variety of examples as the space allows. Hopefully, the discussion has provided an answer to the basic question of this paper. That question, as you may well remember, concerns two individuals facing the same problem and being in identical states of knowledge. Only one of them succeeds in giving a creative answer. The question is what he can but the other cannot do. The answer should be clear by now. The person who finds a creative solution can change the form of the content he attends to and shift his attention between different features of the representation, and he can also use comparisons pertaining to structure or content to guide his reasoning and actions. It seems that he can play around with his representations in a more casual way than the other person and can set his thoughts free from the general or common conceptions that normally guide them.

So far I have supported my thesis by giving examples of creativity which, as I see it, univocally point in the direction of the rules I have set up. Another kind of support comes from linguistics and cognitive psychology. Experiences and detection of similarities obviously play an important role in decision-making and agency, especially in cases when the subject does not have explicit knowledge of the domain in question. When there is a lack of conceptual knowledge, perceptual experience substitutes for it. When we cannot deduce or infer, we reason by similarity or analogy. Higher-level symbolic thought co-operates with lower-level perceptual experience.

Much reasoning is guided by categorisation, by seeing one thing as another. Studies of reasoning and decision-making have often concentrated on logical, inferential thinking and calculation. In many cases, however, the results of these studies appear to have normative, rather than descriptive, import. Reasoning by similarity instead of calculation seems to have both evolutionary and parsimonious advantages and lies at the bottom of more complex reasoning.[8] The ability to categorise is of fundamental value for the

---

[8] Cf. the paper (in Swedish) by N.-E. Sahlin and P. Gärdenfors in *Huvudinnehåll* (eds. Å.E.

simplest train of thought. If a subject cannot identify and reidentify the object he reasons about, then he cannot entertain any continuous, coherent thoughts. He becomes a momentary individual.

Categorisation develops in children at an early age from a general ability to notice different features of objects over context-bound reidentification to abstract categorisation. Abstract categorisation is in principle independent of detection of perceptual similarities and rests upon theoretical knowledge. It appears that conceptual or symbolic thought in this manner evolves from imagery.[9]

Experimental work on categorisation and similarity judgments is relevant for studies of creativity since many of the insights that open up for creative solutions to problems consist in more or less temporary recategorisations. We have already seen examples of this: masking tape happened to be counted as a device used in grafting instead of bast, buildings were conceived of as a natural part of the ecological system and not primarily as cultural artefacts, and so on. Other examples can be found in biology, in the categorisation of whales as mammals instead of fish, or dinosaurs as birds instead of as reptiles, or in physics, where the categorisation of sound oscillates between wave and particle.

Recategorisations are often introduced to make a break with fixed conceptions. The part of categorisation that is of interest to people doing research on creativity pertains, I have argued, to two kinds of situations. First, situations in which the subject has satisfactory information, but does not know how to develop or organise it. The domain stands in need of a transfer of form from another domain with a similar content. Secondly, situations in which the subject has incomplete information and must improvise to fill it in. The transfer is of content from another domain with a similar form. In both these cases, the interesting similarity judgments are not those that are either naturally (psychologically or physiologically) or socially and culturally induced. On the contrary, an ability to escape from conformity and habits is valuable. This does of course not mean that the detection of similarities is completely independent of constraints, neither natural, contextual or task-dependent ones. The link between creativity in a certain domain and extensive knowledge of at least portions of that domain appears to be quite strong.

Categorisation and creative thinking have in common the comparisons, based upon similarity, between domains or instances of domains that lie at the bottom of reasoning. A difference is that in categorisation both source

---

Andersson/N.-E. Sahlin) (1993) Falun: Nya Doxa. The issue has also been touched upon by S. Halldén in *The Strategy of Ignorance* (1986) Uppsala: Thales.

[9] For instance has L. Barsalou recently presented a both interesting and plausible theory of how linguistic symbols arise from perceptual ones in "Flexibility, Structure, and Linguistic Vagary in Concepts: Manifestations of a Compositional System of Perceptual Symbols" in *Theories of Memory* (eds. Collins/Gathercole/Conway/Morris) (1993) London: Erlbaum.

and target domains are known but not so in creative problem-solving. In categorisation, the subject perceives the relevant similarities between the instance and the category, and knows enough about the category (if he can apply it correctly) to be able to explain and justify the categorisation, if not from a general view-point, at least from his own subjective one. But when he tries to retrieve an adequate source for the target of a problem that demands creativity, he relies on incomplete knowledge of that target.[10] A problem requires a creative solution in exactly those situations in which the subject does not have sufficient knowledge to be able to reason inferentially. Instead, he has recourse to similarity judgments in the inductions and conclusions.[11]

A consequence of this is that he cannot, as in normal problem-solving, complete his forward search from target to source with a backward search from source to target and reconstruct the steps in between. Often, when we try to solve a problem, we know not only the question but also the answer we want to get, but we do not know how to produce it. In such cases, a good strategy consists in working both from question to answer and from answer to question. But when the answer lies in the dark, this strategy can of course not be used.

Another consequence of this lack of knowledge accounts for the fact that a fresh similarity can work as an impetus without being true. It can guide the search for a solution and point in the right direction, but still not produce any new inductive knowledge. The guiding similarity has an indirect influence on the solution but is not part of it. This distinction between the direct and indirect influence of similarity judgments links up with a distinction between surface and deep similarity. Surface similarity is perspicious enough to make us examine the relation between two domains that we have no *reason* to believe similar. If we are lucky, we discover a deep similarity between the domains as well. Surface similarity functions as the impetus to a solution, while deep similarity is the result of the investigation that follows upon the creative breakthrough.

Let me compare with grafting again. Initially, a similarity in function is found between bast and masking tape: they can both be used to keep things together and in place. Since no bast is available, tape is used. In this case, the result is beyond all expectation. It turns out that the tape has not only

---

[10] This fact has been underlined by P. Johnson-Laird in "Analogy and the Exercise of Creativity" and S. Vosniadou in "Analogical Reasoning as a Mechanism in Knowledge Acquisition: A Developmental Perspective", both articles in *Similarity and Analogical Reasoning* (eds. S. Vosniadou/A. Ortony) (1989) New York: Cambridge University Press.

[11] E. Smith, E. Shafir, and D. Osherson have shown that inductive inferences made with unfamiliar predicates is based on similarity between the premise and the conclusion categories. When the predicates instead are familiar, judgments of plausibility become pertinent. Plausibility judgments rest upon analyses or decompositions of the familiar predicates. See "Similarity, Plausibility, and Judgments of Probability" in *Cognition*, 49 (1993) 67–96.

similar but better qualities than the bast. The surface similarity in function allowed masking tape to be used instead of bast, but it only indirectly influenced the role the tape came to play. A deeper similarity caused the tape to work so well. If the common adhesive tape had been used instead, or why not yarn, the result would no doubt have been less successful, even though the latter two devices also share a surface similarity with bast. Obviously, there are differences between the bast and the masking tape which are responsible for the superiority of the tape to the bast. But the tape could not have worked so well had it not had some necessary properties in common with the bast, for instance, that it lets the tree 'breathe' and that it is not too tight.

L. Barsalou has pointed to the use of ad hoc categories that are introduced when a subject pursues a novel goal.[12] These categories are temporary and introduced for a special purpose, as when throwing a big party you construct categories like 'dishes that are not expensive and still elegant and tasty' or 'music that is vigorous, but does not disturb the neighbours' to help organise the party. The similarities they rest upon are highly goal-dependent and personal. Ad hoc categories, as well as the metaphorical models that sometimes are introduced in the search for creative solutions, can misrepresent reality but still be of value. Their role consist in providing a guide-line of how to pursue the goal, in being an indirect influence on the solution. They can serve even if they are 'false' by giving an insight into what to do next.[13]

One rather central question has so far been left out of the account. It concerns how the relevant similarities are chosen in situations that require creativity. The choice is not straightforward, because, as just mentioned, the subject has incomplete knowledge of the target domain. Several factors influence it. First, some similarities apparently are perceptually hard-wired—we cannot but help to discover them. Secondly, the process of comparison itself helps select features in the sense that the similarities that are pertinent when I compare A and B do not present themselves when I compare A with C. Thirdly, the task description sets constraints on what features should be relevant in a certain context and thus subject to comparison.[14] Still, it does not seem that these three constraints alone can guarantee a creative solution. It has been suggested that an element of randomness enters into the process.[15] Personally, I am not sure if an appeal to randomness increases our

---

[12] See e.g. "Ad hoc categories" in *Memory & Cognition*, 11 (1983) 211–227.

[13] J. Davidson underlines the role of selective comparison for getting insights in her study of gifted children in "Insight and Giftedness" in *Conceptions of Giftedness* (eds. R. J. Sternberg/J. E. Davidson) (1986) Cambridge: Cambridge University Press.

[14] These three factors are also mentioned in R. Goldstone "The Role of Similarity in Categorisation: Providing a Groundwork", *Cognition*, 52 (1994) 125–157.

[15] P. Johnson-Laird distinguishes between what he calls three main classes of algorithms that can yield a creative product (solution to a problem). The neo-Darwinian one starts with a random

understanding of the selection process. As far as I can see, one of the biggest and most important secrets of creativity lies buried here.

I would like to end this paper by relating to a very traditional view of creativity. That view first and foremost counts artists in different fields as 'truly' creative. There seems to be a grain of truth in it, namely, its reliance on a special metaphorical way of thinking, usually referred to as intuition. Intuition has long been regarded as noncognitive and figurative, as the opposition of deduction and also of hypothesis-framing and experimentation. The rules for manipulating representations that I have brought up here come close to this kind of intuitive thinking. Apparently, the figurative and the literal are not as much opposites as complementary. As some of the examples have shown, even science sometimes relies on the figurative. Recent work on metaphors show how they, conceived of in a very general manner as transfers of features across domains, help structure the information we receive from the world around us.[16] Metaphors lie at the bottom of many of our experiences, for instance of life as going up and down (like a path across a hilly landscape can go up and down) or of time as money or argument as war. I take this as another sign that creativity, or the ability to conceive of the given in a new light, is a fundamental trait of thought, whether conscious or not.

Department of Philosophy
Lund University, Sweden

---

generation, and then uses constraints to filter the products. The neo-Lamarckian one generates the products under constraints, and then selects randomly. The third, multistage algorithm uses constraints both in generating and testing, while the very final selection it makes is random. Johnson-Laird prefers the neo-Lamarckian algorithm, but points out that it is hard to imagine what constraints could be used in the initial generation of products, since these constraints should be common to all problems that require creative solutions. See "Analogy and the Exercise of Creativity" in *Similarity and Analogical Reasoning*.

[16] Important work in this field has been done by G. Lakoff and M. Johnson, among others. Metaphorical thinking can also be described as a superimposition of representations. It is wrong to believe that metaphors only involve a comparison between two entities or concepts. A metaphor usually stretches over several conceptual fields. It does not consist in a one-step metaphorical computation, but in several steps, and thus often rests upon a series of unexpressed, underlying metaphors.

SÖREN HALLDÉN

CREATIVITY AND THE EVOLUTIONARY VIEWPOINT

It seems fitting to speak about a "creative" effort only when the outcome leads to surprise. The step which is taken exceeds the limits of the predictable. Then there is a problem of reconstruction: a clever outsider who tries to find out how it was done may find it very difficult to retrace the path which has been followed.

Important aspects of the problem are studied by the psychologist and the sociologist. But there are other questions which concern the nature of argument: the technique of problem solving is involved. Can one define moves which lead on to the unpredictable, a technique for surprises? We tend to regard problem solving as a logical activity, and the question of the reach of logic arises. Are the marvels of creativity to be explained by ordinary logical theory?

In a rich research field I am going to concentrate on a question which concerns the technique of problem solving. I do not believe that the latter is wholly logical. Instead I shall try to show that it involves something else, a form of empirical know-how.

Consider the intra-representational rules mentioned by Ingar Brinck in her contribution.[1] The rules in question serve a purpose; they do not appear unreasonable. However, their functionality cannot be fully explained by logic; experience justifies them in certain contexts. They are empirical and context-dependent. To see how they work one must, I think, rely on evolutionary thinking.

THE CREATION OF NEW BIOLOGICAL FORMS

Creativity in the biological contect and creativity in problem solving may at first seem completely different. But the difficulties are partly the same, and the solution may be so, too. There are also deeper links.

New biological forms are produced continuously. The process first appears entirely fortuitous; mutations settle the issue, and they strike impredictably. But, by further consideration, it is seen that their range is limited, and that the limitations are significant from the viewpoint of evolution.

If the storehouse of possible new forms is small, the chances of survival, if

---

I want to thank Bertil Mårtensson for helpful comments on the first version of my manuscript.

[1] "The gist of creativity", in the present volume.

Å. E. Andersson and N.-E. Sahlin (eds.), The Complexity of Creativity, 17–22.

the conditions of life are changed, will be small. Complete stability is not an asset; sooner or later it leads to the extinction of the strain. On the other hand, if the store-house contains an a enormous number of possibilities, this will lead to costly experimentation. Suppose that a strain would have a tendency to leave the DNA-model and try something else: this would mean unnecessary waste.

The degree of flexibility will be relevant from the evolutionary viewpoint. One suspects that the development of basic mechanisms will be under the control of higher order mechanisms, which put a stop to certain forms of development, and which have proved their usefulness during previous epochs.

I now turn to the epistemic aspect. Compared to civilized man, the amoeba is ignorant. But already at an early stage the genes of the organism will be connected with a system of behavioural knowledge, involving suitable responses to standard stimuli. There is also a connection between the set of possible genetic variations and the frame of flexibility of the epistemic system in question.

The flexibility in question becomes relevant at several levels: (1) with respect to the stimuli and responses linked up within the system, (2) with respect to the variations which are possible in the nature of the links, and (3) with respect to changes in (1) and (2). (1) defines a field of application, (2) and (3) a set of permissible moves. As in the case of the genes, one expects that flexibility at these levels in the epistemic system is not completely random. If flexibility is small, life in a new niche will be excluded; if it is large, the output of misfits will become expensive.

A well-defined point of departure, modifications which are not determined in advance, but limited in scope . . . this is what one should expect from a system of behavioural knowledge. In these respects it will not be wholy unlike systems of a more advanced nature.

Martin Edman makes use of the following metaphor to characterize explanation in the sciences: "Assume that we are equipped with a box of meccanopieces on which there is a (somewhat equivocal) picture of a lifting-crane. Assume further that our intention is to build a copy of the lifting-crane using the pieces in the box. Aided by the picture on the front of the box and the pieces in the box we will make sketches of how the lifting-crane could possibly be constructed."[2]

The box is a well-defined set of conceptual elements, the picture of the lifting-crane a regulator of the possible modifications in the construction of the explanation. In the standard case the scientist has a playing-field; in this he has to stay; a limited number of moves are open to him. A variability which is partly unpredictable, but not wholly uncontrolled, is essential here.

[2] Page 16 in "There is no naive way of making forecasts", *Knowledge and Concepts in Futures Studies*, ed. S. Schwarz, Westview, Boulder, 1976.

Higher order mechanisms which regulate limitation are, of course, of great interest. Both in science and in biology one expects continuity; small steps are, of course, less dangerous than big leaps. A higher-order mechanism favouring continuity in development will then be an asset. But note that this will be an advantage also for the mechanism. Its own chances of survival increase. Higher order mechanisms are themselves subject to selection in the evolutionary process. Those which lead to rigidity are dangerous and will tend to be eliminated. The same holds for those which trigger off a cascade of untried connections.

Experience is decisive for the development of epistemic systems at all stages. This holds also for the higher order mechanisms which regulate the limits of experiment. These mechanisms can be regarded as tools of problem solution. Their quality will then be dependent upon earlier attempts at problem solution, and the success of these. The outcome is a technique with empirical foundations.

Let us now see if this ties in with more ordinary ideas about creativity. Too illustrate this, I shall use two examples.

WHEN PLAYFULNESS PAYS OFF

Let me start with an example from my own experience. Long ago I was working on a dissertation in mathematical logic. I then learned that this is not at all a field where one can rely on a mechanical rationality. If a problem looked promising, one could be certain that it also looked insoluble. To survive I had to accustom myself to a set of rules.

Rule 1. Be relaxed.

I soon discovered that, if I tried resolutely to go straight at the solution, I would find nothing. The problem looked hard for a simple reason: it was beyond conventional problem solving. Relaxation would be limited: an inner disquietude would keep me going.

Rule 2. Keep to the relevant items.

I would start doodling in a somewhat silly way, but I would choose material of a certain kind. Perhaps a formula which I wanted either to prove or to disprove. Or a formula designated either by optimism, or by pessimism. A playing-field was presupposed, and I had to stick to this. I shall call this the "rule of the field".

Rule 3. Rearrange the items before you in a haphazard order.

But the order would not really be haphazard. I would use certain devices which I had learnt earlier. With a given formula before me, I would reformulate it in various ways. I would do this to find out more about its internal structure. I would also deduce other formulas from it. I could also try to imagine negative circumstances, and define a counter-example. Or I would rewrite it in slightly simpler form, and try to see what that would lead to. Or I would try to see what the formula would imply in a well-defined situation of some sort. Such were the moves open to me. I shall call this the "rule of the moves".

Rule 4. Be ready to reformulate the problem.

I had learnt that the conventional formulation of a problem often involves a misunderstanding of some sort. In an intelligence test one cannot question the assignment one has been given. But free research is something else.

Rules 2–4 are "intra-representational" rules in the sense of Brinck.

Rule 5. Be open for links to what you already know.

This is not as obvious, as it first seems. The rule should be interpreted in a special way. It does not recommend a direct attack on the target; it recommends an attitude of vigilance. If I could find an opening (perhaps in the form of a suggestive structure), I should be ready to go in.

Rule 6. If your efforts are fruitless, take a rest, and leave the problem to the unconscious.

The last was a rule I often had occasion to follow, and I did so with a feeling of hopefulness, confident that the brain would set to work. I knew that it would be wise to let it be undisturbed.

I suspect that the brain has its own methodology, partly of biological origin. But the rules which I defined for myself, what basis did they have? Their epistemic origin is of interest. I learned them, when I tried my hand at problem solution. I was guided by very simple logical insights, concerning the possibilities of discovery. But logic could not tell me anything about the probabilities involved, and the chances of success. To find the right strategy, I had to have past results in mind.

The rules were derived from experience, an experience of a quite special kind. But no ordinary testing procedure was involved; the insights derived were the results of trial and error.

WHEN ONE KNOWS VERY LITTLE

A scientist of the type depicted by Edman starts from something definite; his moves are guided; he has something to rely on in his work. But there are other situations in which one knows very little in advance, and when one's first efforts will look very humble. Playfulness will then be a necessity; conscientiousness will have almost nothing to lean on.

The more desperate an enquiry is, the more modestly it has to start. Writing this, I am thinking of the priests of Asclepius on the island Kos at the time of Socrates, and the spirit of inevitable perplexity in which they had to deal with their insane patients. We do not know very much about insanity today; they were groping in the dark. What did the wrath of the Gods mean, what the influence of the heavenly bodies, what one's personal circumstances, what the weather, what the food?

The climate of creativity in early psychiatry (chiefly Greek) has been analysed by Bertil Mårtensson in a suggestive study.[3] The exertions in the field were certainly done with a feeling of the enormous seriousness of the diseases studied, but nonetheless one had to proceed in way which afterwards has a tinge of arbitrariness. The style of inquiry is not wholly unlike that described in the last section.

Mårtensson uses the term "theme" when he describes the intellectual elements, the concepts and hypotheses, which were used to construct the explanation of various forms of insanity. The chief characteristic of the theme is its variability. When the investigator finds that it does not come up to expectation he may make a change of some sort, add a new requirement, make a switch in causal order, combine it with a new theme.

Two aspects of this are worth attention: the need for freedom when one is trying to find a good explanation, and also the need to impose limits on this freedom. If the truth-seeker had the right to suggest absolutely anything, freedom would be too great; if he was limited to the explanations of the past generation, he would be too severely restricted. The "themes" mentioned by Mårtensson were key expressions which commanded a certain respect, but were open to gradual modification. The limits will lie in the fact that one is restricted to the variation and recombination of standard elements.

The two argumentative rules mentioned in the last section, those of the field and the moves, may be discerned here. The field consists of elements which might be cited when the explanation is presented; the moves are those which suggest themselves when an explanation has failed and has to be replaced by a new one.

Perplexed and concerned, these early doctors keenly felt the need for a set of concepts by which they could describe and explain the phenomena and conditions of sanity and insanity. Different forms of disease were delimited in provisional ways; the definitions were modified in an equally groping manner. This was an attempt to define a field.

They also had to find out about the causal connections within the field. In what way could they alleviate the sufferings of the stricken? Mårtensson

[3] "Thematic Structures". Manuscript.

brings out the function of transformation rules by this undertaking. If A does not explain B, perhaps a variation of A explains B, or one can try to explain B by A, or some special kind of B by A. Analogies were brought in, permitting different kinds of solution. If the presence of something did not seem clearly relevant, one could try to find out what the absence meant. The doctors studied the coincidence of phenomena, and drew conclusions about causal connections. These were the moves selected by them.

The efficiency of such a study in an undeveloped science of the sort investigated is limited by the simple fact that one cannot regard the definitions of the A's and the B's as definite. They are incomplete, and they are open to modification.

The general style of investigation is reasonable, if one regards it from the viewpoint of common sense. The pattern applies when certain phenomena occur together, and when one wants to see the connections. This way of thinking is rooted not only in experience, but also in pragmatic considerations. Something might happen again, and then one wants to be prepared. However, there is no backing in logic. If A and B occur together, they may be totally and completely unconnected.

Logic has very little to do with it, but this does not refrain the investigator from the use of the approach. Psychiatric thinking in early Greek society gives dramatic expression to a way of thinking which might well have deep foundations. Faced with the incomprehensible, or with danger, people have taken their refuge in this somewhat desperate form of truth-seeking.

Experience and the practical necessity of the situation provide the keys, together with an element of trial and error. The evolutionary viewpoint once again proves fruitful.

Department of Philosophy
Lund University, Sweden

GUDMUND SMITH

THE INTERNAL BREEDING-GROUND OF CREATIVITY

This presentation will be more concerned with creativity than with cognition in the usual import of that word. I believe that we need a lot of creativity to fertilize cognitive psychology and make it the radical pioneer it should be, and often pretends to be, in the science of the mind. It is only fitting, then, that I begin with some critical reflections.

ASEPTIC COGNITION

Cognitive psychology is normally defined as the study of thinking, memorizing, information processing, choice behavior, etc.

Many of the early writings in the field did not make quite clear if the experimental subject was assumed to be fully aware of his cognitive operations. The general impression was, however, that cognition and awareness were associated as a matter of course. Now, admittedly, all that has changed. Preconscious processing has become part and parcel of the picture, almost like a fashion article.

But for many old-fashioned cognitivists there is still no qualitative difference between "within" and "beyond" awareness. Even to a comparatively modern cognitivist like Loftus (1992), the main question appears to be how efficient preconscious processing can be compared to conscious processing. As one of my New York friends expressed it: The difference is like using the computer with the screen lighting switched on or off.

One reason for cognitive psychology's relative neglect of the importance of consciousness qua non-consciousness seems to be computational functionalism. In the early days the computer appeared to offer a definite solution of the body-mind problem and speculations about consciousness hence became superfluous. As you will soon understand, to tie cognition to the computer is not wholesome for a science of creativity.

Another problem is that cognition has often been viewed in an aseptic light. What I mean is, for instance, that emotion and affect have seldom entered the equations of cognitivists as fundamental determinants. Emotions might surface at the other end of the cognitive process, e.g., as a consequence of decisions with unfortunate outcomes. For many students of creativity the opposite would seem more likely: since meaning precedes conscious articulation in the perceptual process, as outlined later, the emotions and affects intrinsic to meaning would be the most probable starting-points for cognitive thinking

Å. E. Andersson and N.-E. Sahlin (eds.), *The Complexity of Creativity*, 23–33.

and action, i.e., its basic motivational factors. This is still a controversial issue (cf. Smith & van der Meer, 1994).

The aseptic attitude is associated with the tendency of puristic cognitivists to accept only deductive conclusions; inductive ones are considered as house-trained only in case the probability of generalizations is very high. A third way of thinking, abduction, was suggested by the American philosopher Charles Peirce. Abduction could be interpreted as guessing or, rather, an ability to accept surprises. This alternative would be unacceptable to a puristic cognitivist. But abduction seems to be the way creative people like to think, at least as an introductory strategy.

CHAOS AND CREATIVITY

Aseptic cognition has thus little to do with being creative. It is as distant from creative functioning as compulsive people are; or, the alexithymic ones, those who are unable to find words for their feelings and emotions. Parenthetically, it is interesting to note that in our studies compulsives and alexithymics belong to the least creative groups.

It has often been suggested that creative ideas grow out of chaos. Shaw (1994) thinks that "...creativity is far from simply a cognitive process. It involves human processes that have deep feelings, both positive and negative, associated with them" (p. 40). Krystal and Krystal (1994) take the bull by the horns more directly: "Systems that are capable of 'chaos' have the ability to produce novel and unpredictable responses, and are thus capable of conferring the capacity for human creativity" (p. 186).

These authors also talk about "chaotic", non-linear systems that are "capable of generating novel, unpredictable behavior and are, therefore, useful models of creativity" (p. 196). In this context they refer to the Lund methodology—to be described presently—as one allowing "those creatively capable to exercise their imagination according to their inclination" (p. 206), or in an abductive way as Peirce would term it. However, the term "chaotic" would be unfortunate if it was understood as analogous to random. Better synonyms could be ambiguous or inconsistent to an outsider. The non-linearity in our model may be seemingly chaotic but is still lawfully patterned from an insider's perspective.

A PROCESS APPROACH

My present discussion of creativity is based on the general micro-genetic assumption that the individual's view of reality is shaped from inside and out, i.e., by ultrashort processes the beginnings of which, like dreams, are stamped by seemingly absurd but deeply personal connections between its ingredients. These processes have both a cognitive and an emotional side to them. During a micro-process aspects deviating from approximate outside reality are gradually stripped off in favor of a better fit. Our group in Lund has been mainly concerned with visual perception. Since we have arranged our experiments accordingly we prefer the term percept-genesis to that of micro-genesis. Changes over the percept-genesis are qualitative, i.e., nonlinear. This has been repeatedly demonstrated experimentally by means of artificial prolongation of perceptual micro-processes. As will soon be demonstrated, the tachistoscope has been a prominent tool in these experiments (cf. Westerlundh & Smith, 1983).

It is important for creative persons to be open to the nonlinearities of their inner world, to be tolerant of incongruities, novelties, strange combinations, etc. (cf. Krystal & Krystal, 1994), and of the anxieties triggered by them. To us creativity is a way to combine antagonistic impulses, e.g. masculinity and femininity, often at a symbolic level.

These antagonistic impulses need not confront us directly in awareness. Perhaps more often as a feeling of alienation, of not really belonging anywhere, of finding oneself in-between, of not being tempted to follow all the others in the mainstream. Even a feeling of bewilderment can be a reflection of the chaotic roots of percept-geneses. People who let themselves be calmly carried away by the mainstream have no use for creative ideas. The navigation takes care of itself according to plans drawn up beforehand—by others. Gardner and Wolf (1988) pointed out how both Picasso and Freud felt marginal for much of their lives. "Presumably these feelings of marginality served to motivate our two creative heroes to 'prove themselves'" (p. 106).

THE EMPIRICAL STUDY OF CREATIVITY

In the empirical study of creative functioning the use of tests is a necessary option. But tests would be of little value if they had not been validated against acceptable criteria. Looking around for useful criteria one is naturally drawn to artistic products, even if the standards applied have to be relatively modest. Children's drawings, for instance, do indeed vary with respect to degree of originality and emotional engagement. A test, the results of which correlate with such qualities in the drawings, can be used as a measure of creativity (Smith & Carlsson, 1983). The advantage of the test as compared with the drawings themselves could be a more simplistic scoring and greater applicability in groups where a drawing competition is difficult to arrange or openly resisted.

It should be pointed out, however, that not even widely appreciated artistic products guarantee a highly creative originator. A special study of professional artists (Smith, Carlsson, & Sandström, 1985) convinced the experimenters that the artistic profession was not necessarily associated with creative production. It may suffice for the artist to be a skilful craftsman and adaptively imitate what is currently marketable. And a recent study by Schoon (1992) of English students in architecture demonstrated a highly variable degree of creativity in her supposedly creative subjects, even lack of creativity.

Many practitioners in the so-called creative professions are rather recreative than creative. We do not only find them in the visual arts but, also, perhaps more often, in the world of music. And why not include the intellectual world in this critical perspective? A good scientist need not be particularly creative in order to be highly appreciated among his peers. Intelligence, wide reading, and methodological and technical skills can speed up the career considerably even if these cognitive advantages are not associated with a generative fantasy. A too bounteous imagination may even be a disadvantage for a tidy research existence.

Creativity tests have been in use since the early 1950s. A complete survey of them is presented in a recent Handbook of Creativity Research (Runco, 1996). Our own Creative Functioning Test, CFT (Smith & Carlsson, 1990) implies that the subject is confronted with a series of gradually prolonged presentations of one and the same stimulus (a still-life) until a correct description has been delivered.

The psychometric line	**The psychodynamic line**
A. Binet (divergent thinking)	S. Freud (reluctant)
J.P. Guilford (creativity as intelligence factor)	C.G. Jung (the shadow)
E.P. Torrance (creativity tests)	E. Kris (adaptive regression)
	S. Arieti (magic synthesis)

Fig 1. *Names in the history of creativity research*

This is a way of reconstructing a percept-genesis. What we are particularly looking for at the creative end of the test scores are subjects not necessarily tied to this "correct" interpretation of the stimulus when, in the latter part of the test, the foundation of their perception is gradually eroded by a systematic abbreviation of exposure times. Creative subjects do not react to the diminishing clarity by almost desperately clinging to the interpretation they have chosen as the correct and secure one, but eventually supplement it with their own, idiosyncratic versions. Measured in this way creativity has been shown to correlate with various external criteria like creative products, but also with a more general creative stance, a yearning to create something of one's own, fantasy, awe when brought to face the wonders of life, a need of autonomy, etc. (see, e.g. Smith & Carlsson, 1990; Carlsson, 1992; Schoon, 1992).

A HISTORICAL REVIEW

If we briefly illustrate classical creativity research with the two columns of names shown in Fig. 1 our attempts place themselves in the middle. Alfred Binet (Binet & Henry, 1896) coined the term divergent thinking to account

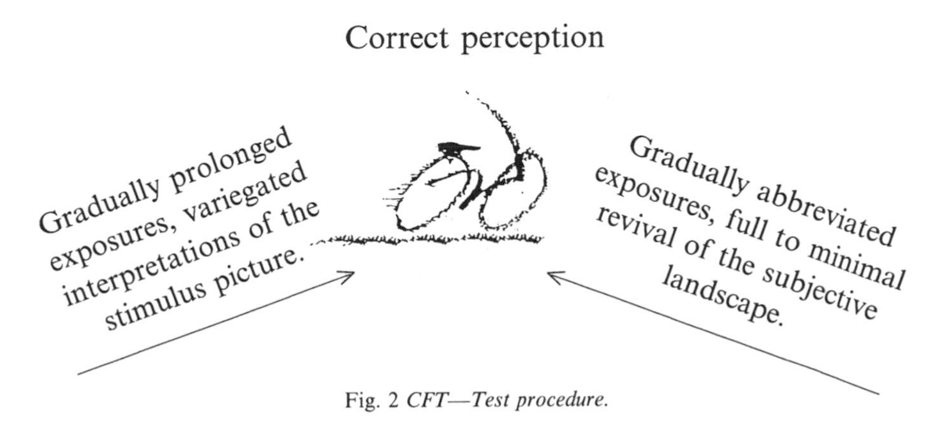

Fig. 2 *CFT—Test procedure.*

for the manner in which creative people try to grapple with cognitive tasks. J.P. Guilford (1967) used the then relatively novel technique of factor analysis to give creativity a space in the intellectual universe. To define creative behavior he asked his subjects what uses they could think of for a brick or a newspaper, besides the uses intended for them. At about the same time E.P. Torrance (1966) introduced creativity testing in education.

Sigmund Freud was not particularly inclined to formulate a theory of creative functioning. Instead, his one-time follower and later rival C. G. Jung (see Jung et al., 1988) revelled in formulations on the subject. To him it was important for the creating individual not to shy away from the dark side of his personality, the shadow, because that is where he can find his creative impulses. Ernst Kris (1952) speculated in a similar vein about a self-regulated regression to preconscious states in the inspirational phase of the creative process.

While the psychometricians to the left in Fig. 1 were sophisticated methodologically they were theoretically dull, mostly relying on trait models or typologies. The psychodynamic writers seemed much more interesting when focusing on creativity as a process and looking for its ultimate and mostly hidden determinants. They also emphasized the role of symbolization in creative work. The world of symbols is where contradictions can be most easily resolved. But when relying on biographical material they easily made unwarranted generalizations.

While theoretically leaning to the psychodynamic side our group in Lund has made the use of experimental methods a "must". We have hesitated, however, to apply the classical creativity tests because their criterion correlations, even if generally positive, have been unimpressive. In that respect our percept-genetic methodology has been more successful. Nevertheless, any experimental method should never be allowed to stand alone. Professionally performed interviews, for instance, could always serve as a useful complement.

IS CREATIVITY A UNITARY CONSTRUCT?

After all this talk about creativity it is time to close in on the question whether all creative people should be lumped together in one unitary category. Already 40 years ago Ghiselin (1952) argued that "The creative process is not only the concern of specialists... not limited to the arts and to thought, but as wide as life" (p. 24). Richards (1990) distinguishes between everyday and prominent creativity. For her everyday creativity is broadly applicable to work and play or virtually any human activity. But there is no difference, in principle, between the two.

If eminent creativity is usually identified with a creativity that leads to visual and audible products, what is, then, creativity that does not? Is it on the whole possible to talk about creation that does not create anything obvious? But why could we not regard creativity as a fresh approach to life, a way of comprehending oneself and others, an urge to construct something outgoing from one's own prerequisites—regardless of the traces this will leave for the future? It could be mere fantasy, an original perspective on the world and its inhabitants, never formally mediated as a text, a painting, or piece of music, a discovery never made public. The most important focus of a creative attitude need not be a creative product in the usual import of that concept but could rather, to borrow a central idea in Michel Foucault's late thinking, be the invention or improvisation of who oneself may be (see Nehamas, 1993). There is nothing to prevent us from calling such a self-actualizing person creative, provided that his ideas are not just copies of what is presently fashionable.

Apropos traces left for the future, I would like, parenthetically, to interpose that the creative product is often defined in a very conventional way, as a lasting remnant: a painting, a song, a poem, an invention, a theory, a political achievement, etc. But the tale of an oral narrator remains only as fragments in the memory of the listeners. And the ability of a therapist to understand the nature of a patient's problems in a new way is seldom documented but in the form of rapsodic weekly notes.

Often the creative attitude is not noticed by others, at least not by casual acquaintances. The pupil who asserts herself most loudly in class, who appears as spokesman for her classmates and organizes their games and plays, is not necessarily the most creative. Perhaps we find him among the withdrawn, those who prefer to stand by themselves. Not until she returns to her own room does she show her creative countenance: as drawings tucked to the walls, poems in the desk drawers, outlines of plans for a new society, constructive fantasies never entrusted to anyone, or just attempts to define one's own existence.

Everyday creativity is a necessary assumption for the test psychologist. But it is also of great practical importance. People identified within this category are those who perhaps do not generate (or publish) revolutionary new ideas but are capable of understanding and appreciating them. We identify them as people who are not irreversibly caught in the steel-nets of conventional thought. One characteristic reflected in many of our studies is their unwillingness to be manipulated, a quality that sometimes makes them unnecessarily remonstrating but mainly mirrors how much they value their autonomy. Without these people the eminently creative ones would have no sounding-board for their proposals.

PSYCHOPATHOLOGY AND CREATIVITY

It was demonstrated by Becker (1978) that outstanding creative individuals are often afflicted with various psychopathologies. A recent study by Post (1994) of the biographies of 291 deceased scientists, composers, politicians, thinkers, artists, and writers partly confirms Becker's results, but particularly with regard to artists and writers. With few exceptions, says Post, the other subjects were emotionally warm, with a gift for friendship and sociability, even if most of them had unusual personality characteristics.

This interesting difference between writers and artists, on the one hand, and other creative people, on the other, could possibly be comprehended within the percept-genetic model and has been demonstrated in several studies by us and others (Smith & Carlsson, 1990; Schoon, 1992, etc). We assume that in both groups the percept-geneses are open to their ultimate roots. This openness is made possible by their general tolerance of unease. In scientists the emerging ideas are all the time played against the fence of previously accepted knowledge. Idiosyncratic nonsense will not do. In artists and writers no such controls are universally applied: On the contrary, it is part of their trade to make the subjective side of their cognitive processes public.

THE CREATIVE PERSON

What are, then, the basic common characteristics of creative persons, both those who display their creativity in tangible products and those who do not manifest it outwardly but still share a creative approach with the former? Many qualities could be enumerated (cf. Smith & Carlsson, 1990). But they all seem to depend, as already said, on the open communication inwards, between surface and depth or, as Norman Dixon, the guru of research in subliminal perception, would express it, between the conscious mind and the subconscious brain (Dixon, 1981). Too much conscious deliberation is likely to destroy creative ideation.

If one prefers a historical perspective instead of a spatial metaphor one might talk about an urge to reconstruct the past within the frame of the present. Regression in the service of the ego, the phrase coined by Kris (1952), may come to mind. But the communication implied here need not mean more than momentary slackening of ego control and may very well be intentional (cf. Rothenberg, 1979).

The creative person is generally characterized by a "long" memory, often back to the age of two years (in some few subjects even before that age). What he remembers is seldom vague but specific, sensual: the warmth on sunlit doorsteps, the particular feeling of beach sand between your toes, a green spot of grass in front of a house. This might easily sound as childhood nostalgia. But in most memory stocks one discerns both light and shadow, happiness and distress, gains and losses. Often it seems like such contrasts were essential incentives for a creative process to start.

In parallel with the long historical perspective many creative people cultivate an interest in their dreaming. They remember their dreams, know unhesitatingly if they have been in color or black-and-white, and understand in some way or other their symbolic message. On the whole, as I have already intimated, these people have an aptitude for conflict resolution at a symbolic level. When, in a special test situation constructed to map defensive strategies, they are confronted with a visual threat, they easily avoid a direct confrontation by transforming the threat to something entirely different, a tree, a landscape, a building.

CHILDISHNESS IN CREATIVE PEOPLE

Growing adult very often means losing your creativity. This loss is evident already in early middle age (Smith & van der Meer, 1990). Still, many people remain creative even in old age. There is ample room for speculation about these contradictory trends. Without excluding psycho-social determinants I feel tempted to borrow a biological representation of these developments. Modern neurology has introduced parcellation as a key concept to explain what happens to the aging brain. Parcellation has been defined by Jason Brown (1992) as "the pruning of exuberant connections in the growth of the brain as a way of achieving specificity in mature brain structure".

The progressive thinning out is a sign of increasing maturity and specialization. But there may still be a temporal disparity between the neural systems (heterochrony). Neoteny is a form of heterochrony in which, says Brown, a retardation of development prolongs the duration of a juvenile state. It can imply an escape from the more rigid specializations of adult structure. Neoteny thus corresponds to preserved childishness, at least to some degree, and to adult creativity.

If we return to our own empirical material we find support there for the assumption that the creative adult person harbours a child within herself. We have, in a so-called Identification Test, confronted creative subjects with brief glimpses of a diffuse face (Smith, Carlsson, & Andersson, 1989; Smith & van der Meer, 1990). This arrangement forces the viewer to mobilize her own private experiences in order to be able to describe the face. Creative people significantly more often than uncreative ones report impressions of both a young, even childish, face and an adult one. These people are definitely not children or infantile adults, but they act dialectically with the distinct memories of their own childhood. They look upon the world with fresh, unconventional childish eyes but correct what they see with more knowledgeable adult eyes.

A CREATIVE MILIEU?

I have dwelt on the inner determinants of creativity because psychology, not sociology, is my professional field. Many guesses have been made about the outer roots, but they have mostly remained guesses. One perennial question has been if Vienna, at the turn of the century, was a particularly creative milieu because the old society was coming apart at the seams and new, often contradictory influences were mingling in the intellectual atmosphere. The transition nowadays between the industrial and post-industrial eras could, likewise, be a fertile breeding-ground for new ideas. Even if such speculations seem probable they can never be proven beyond doubt. More successful presumptions have been made with respect to the greater or lesser tolerance of creative behavior in industrial settings, schools, universities, etc. Bigness and overpowering bureaucracy seem to be generally counterproductive.

Among hypothetically inspiring factors we find cultural pluralism, tension, competition, even anxiety inducing stimulation. In one of our laboratory experiments (Smith & Danielsson, 1979) we could even demonstrate that threatening subliminal stimulation enhanced creative production in the latently creative part of the sample. Highly creative subjects were already close to the ceiling; totally uncreative ones were generally unresponsive. It has also been shown that you can teach people to listen more carefully to their own inner ideas and thus enhance their creativity. Alluding to what was said before I may conjecture that those milieus are particularly favorable for creative functioning where the individual does not feel quite at home. To have a wonderful time, to feel warmth and closeness, to fit in 100%, may be good in itself but easily subdues the creative impetus. It has even been proven that youngsters who had been very close to their mothers in early childhood scored lower on creativity tests than children whose mothers kept more distance between themselves and the child (Michel & Dudek, 1991).

One thing is obvious to a psychologist. The environmental influence goes via an internalization of conflicts and problems. A problem does not in itself start the creative working in anyone, only in those who incorporate it as part of their own self. What is not experienced as my own personal problem can never be the starting-point for a creative process. "We are not creative about problems and issues we do not understand" (Runco, 1994, p. 114).

Department of Psychology
Lund University, Sweden

REFERENCES

Becker, G. (1978). *The mad genius controversy. A study in the sociology of deviance*: Beverly Hills: Sage.

Binet, A. & Henri, V. (1896). La psychologie individuelle. *L'Année Psychologique*, *2*, 411–465.

Brown, J. (1992). Morphogenesis and mental process. Invited lecture. Department of Psychology: Lund University.

Carlsson, I. (1992). *The creative personality: Hemispheric variation and sex differences in defence mechanisms related to creativity*. Lund: Lund University, Department of Psychology.

Dixon, N.F. (1981). *Preconscious processing*. New York: Wiley.

Gardner, H. & Wolf, C. (1988). The fruits of asynchrony: A psychological examination of creativity. *Adolescent Psychiatry*, *15*, 96–120.

Ghiselin, B. (ed.), (1952). *The creative process*. New York: Mentor.

Guilford, J.P. (1967). *The nature of human intelligence*. New York: McGraw-Hill.

Jung, C.G., von Frantz, M.-L., Henderson, J.L., Jacobs, I., & Jaffe, A. (eds), (1988). *Man and his symbols*. New York: Anchor Press.

Kris, E. (1952). *Psychoanalytic explorations in art*. New York: International Universities Press.

Krystal, H. & Krystal, A.D. (1994). Psychoanalysis and neuroscience in relationship to dreams and creativity. In M.P. Shaw & M.A. Runco (eds), *Creativity and affect*. Norwood, N.J.: Ablex, 185–212.

Loftus, E.S. (1992). Is the unconscious smart or dumb? *American Psychologist*, *47*, 761–765.

Michel, M. & Dudek, S.Z. (1991). Mother-child relationships and creativity. *Creativity Research Journal*, *4*, 281–286.

Nehamas, A. (1993). The examined life of Michel Foucault. *The New Republic*, February 15, 27–36.

Post, F. (1994). Creativity and psychopathology. A study of 291 world-famous men. *British Journal of Psychiatry*, *165*, 22–34.

Richards, R. (1990). Everyday creativity, eminent creativity, and health: "Afterview" for CRJ special issue on creativity and health. *Creativity Research Journal*, *3*, 300–326.

Rothenberg, A. (1979). *The emerging goddess: The creative process in art, science, and other fields*. Chicago, Ill.: University of Chicago Press.

Runco, M.A. (1994). Creativity and its discontents. In M.P. Shaw & M.A. Runco (eds), *Creativity and affect*. Norwood, N.J.: Ablex, 102–123.

Runco, M.A. (ed.), (1996). *Handbook of creativity research*. Creskill, N.J.: Hampton Press.

Schoon, I. (1992). *On the psychology of creative achievement in architecture*. Leiden: University of Leiden Press.

Shaw, M.P. (1994). Affective components of scientific creativity. In M.P. Shaw & M.A. Runco (eds), *Creativity and affect*. Norwood, N.J.: Ablex, 3–43.

Smith, G.J.W. & Carlsson, I. (1983). Creativity in early and middle school years. *International Journal of Behavioral Development*, *6*, 167–195.

Smith, G.J.W. & Carlsson, I. (1990). The creative process. *Psychological Issues, Monograph 57*. Madison, Ct: International Universities Press.

Smith, G.J.W., Carlsson, I., & Andersson, G. (1989). Creativity and the subliminal manipulation of projected self-images. *Creativity Research Journal*, *2*, 1–15.

Smith, G.J.W., Carlsson, I., & Sandström, S. (1985). Artists and artistic creativity. *Psychological Research Bulletin, Lund University*, *25*, 9–10.

Smith, G.J.W. & Danielsson, A. (1979). The influence of anxiety on the urge for aesthetic creation: An experimental study utilizing subliminal stimulation and a percept-genetic technique. *Psychological Research Bulletin, Lund University*, *19*, 3–4.

Smith, G.J.W. & van der Meer, G. (1990). Creativity in old age. *Creativity Research Journal*, *4*, 249–264.

Smith, G.J.W. & van der Meer, G. (1994). Generative sources of creative functioning. In M.P. Shaw & M.A. Runco (eds), *Creativity and affect*. Norwood, N.J.: Ablex, 147–167.

Torrance, E.P. (1966). *Torrance tests of creative thinking: Norms—technical manual*. Princeton, N.J.: Personnel Press.

Westerlundh, B. & Smith, G. (1983). Perceptgenesis and the psychodynamics of perception. *Psychoanalysis and contemporary thought*, *6*, 597–640.

JASON W. BROWN

PROCESS AND CREATION[1]

The Gifted have told us for years that they want to be loved
For what they are, that they, in whatever fullness is theirs,
Are perishable in twilight, just like us.

— Mark Strand

From the different perspectives that have been brought to bear on the study of the creative personality over the past century on the effects of cultural and political conditions, family structure, early development, and so on,[2] it is clear that a complex of events must converge at successive stages in the life of an individual to nurture creative ability. In addition to this complex of external conditions, a set of innate dispositions is probably no less essential. The balance of these factors determines not only whether an individual is creative but the scope, the drive, the intensity, the confidence to follow through and the discipline that a productive creativity demands.

In discussions of creativity, the life and times of the creative person have been given so much attention that they tend to displace an account of the creative *process* from the mind of the individual where it belongs to circumstances of biographical detail. Collectively, these events impact on the development of the creative *personality* but separately they are incidental to the process of creative *thinking*. An explanation centered in the facts of an individual life leaves the creative process itself unexplained.[3]

In recent years, attempts have been made to demystify the act of creative thinking, especially inspiration, as a phenomenon of almost magical signification. Experimental studies have dispelled the mystique of creativity as an irrational mode of cognition, linking it instead to normal problem solving.[4] The incremental nature of creative thinking and its continuity of expression

[1] Presentation, Conference on Creativity, Institute for Future Studies, Venice, Italy, October 1994. Also published in Brown, J.W. (1996) *Time, Will, and Mental Process*, Plenum, N.Y.

[2] For example, from Lombroso, C. (1891) *The Man of Genius*, Walter Scott, London, to Gardner, H. (1993) *The Creators of the Modern Era*, Basic Books, New York.

[3] Related to this distinction is that of the genius as an explanation of historical events or a product of historical forces. See James, W. (1896) Great men and their environment. In: *The Will to Believe*, Longmans, Green and Co., New York.

[4] See Finke, R., Ward, T. and Smith, S. (1992) *Creative Cognition*, Bradford, MIT Press, Cambridge.

Å. E. Andersson and N.-E. Sahlin (eds.), The Complexity of Creativity, 35–50.

over time suggest that it is related to the normal thought process.

Creativity is certainly present in all people to varying degrees. The sudden insight to the solution of a game or a puzzle, the Aha experience on apprehending the answer to a challenging problem, are not equivalent to the discovery of relativity theory but such experiences presumably reflect a continuum of performance in a common mental process. Have we not all had a fragment of the experience of a Mozart on leaving a concert and "hearing" the music all at once in our mind? For Mozart, a musical concept was generated autonomously while in listening to Mozart, a memory image of the music is generated by a template. As incomprehensible as the genius of a Mozart may be, it is probably an elaboration of certain aspects or domains of normal thought. But what exactly is normal thinking? Is the idea of a common mechanism for the normal and the exceptional—both the gifted and the abnormal—consistent with degrees of originality? More deeply, what is the nature of creative thinking and how does creativity in mind relate to novelty or creativity in physical brain process?

I

Novelty

The distinction between novelty and creativity is often cast as a difference between the physical and the mental, novelty being the appearance of a new entity, creativity a mental act in which the novel is created. We tend to attribute the possibility of novelty to purely material events, though in everyday discourse novelty is applied to human or animal behavior and can refer to both physical and mental events.

On the other hand, creativity requires a mind to bring something novel into existence the novelty of which is judged by others to reflect the talent, originality, giftedness or genius in the creative act. Creativity is not ordinarily assigned to the material world, i.e. the world "machine", except for accounts of the "creation" of the universe, in which case the attribution of creativity follows on the assumption of a mind, i.e. a creator, that is responsible for the creation. For example, in the metaphysics of Whitehead, change in the world is conceived as a creative advance into novelty. Whitehead wrote, "the creativity of the world is the throbbing emotion of the past hurling itself into a new transcendent fact".[5]

In the relation to material events, novelty could be a property of the brain state which is a physical process, while in the relation to mental events, creativity could be a property of the mental state. The distinction entails an

---

[5] Whitehead, A.N. (1933) *Adventures of Ideas*, Macmillan, New York, p. 227; Also, Henri Bergson [(1913) *Creative Evolution*, Holt, New York] for whom God is creation or a creative principle.

implicit dualism, i.e. novelty in the brain state, creativity in the mental state, unless creativity is understood as a "higher" order or evolved form of novelty, in which case the *brain* process that generates a creative idea is no less creative than the mental process to which it corresponds. On this way of thinking, novelty could occur in the absence of creativity, with creativity an expansion of novelty at a certain level of organization, i.e. the creative is a more complex expression of the novel.

A judgment of novelty requires a comparison of a prior and an occurrent state. There must be a departure from the expectations of the prior state for the judgment to be made. Apart from the mind-dependence of this judgment, we can ask what is the nature of the departure. If the step leading to novelty reflects a set of contingent events that impacts on change, the shift from a causal effect to a novel outcome could in principle be specified, say through a *post hoc* analysis of the effects of the (known) contingencies.

In contrast, the creative seems inexplicable from its prior states. There is a causal "gap" between the antecedents of the creative step and the moment of creation. Partly, this feeling of a gap reflects the definition of creativity as a coming into existence. The feeling is heightened by the basis of creative work in the mind, in contrast to the "causal certainty" of physical (brain) mechanism. The creative outcome is further obscured by its presumptive origins in the subconscious with a realization in consciousness, a step more like a quantal jump than a continuous sequence.

We tend to think of physical change as causal, with novelty improbable or rare, like a miracle in a deterministic world. A deviation from the expected is impossible in a world of universal causation. Whether or not there is novelty depends on a theory of change. A novel change should be unpredictable but the inability to predict a change, even if all information is available, does not obligate that the change is novel. Random or unpredictable change is not necessarily an occasion of novelty; a random process could generate a (non-novel) recurrence. An approach on the basis of probabilities has the same drawback. Neither approach gets at the inner *nature* of change. The indeterminism of microphysics is compatible with novelty even if it does not elucidate the *process* through which novel states appear. John Dewey wrote that only a philosophy of "genuine indeterminism, and of change which is real and intrinsic gives significance to individuality. It alone justifies struggle in creative activity and gives opportunity for the emergence of the genuinely new".[6]

Creativity

Creativity is not an accumulation of elements into more complex aggregates. Suppose I imagine or draw the head of a frog on the body of an elephant,

[6] Dewey, J. (1965) Time and its mysteries. In: Browning, D. (ed.) *Philosophers of Process*, Random House, New York, p. 211. The same idea is found in Bergson, H. op. cit.

and suppose also that no-one has drawn or even imaged this combination before, this would not qualify the image or drawing as necessarily creative. The parts are unchanged by their recombination. They are the same parts differently put together. One might as well have a rockpile that is rearranged by an earthquake.

One can say that creativity is novelty applied to concepts; or that in creative thinking there is a conceptual basis for the novel. Essentially, creative activity is the forming of new concepts in the mind of an individual. In concept formation the creative is achieved as a cognitive whole not decomposible to a set of antecedent or constituent elements. Indeed, there are no basic elements, only the emergence of the whole from its ingredients, i.e. the replacement of prior elements by the novel whole, and the potential of the whole to fractionate into novel parts.

While the understanding of the whole may—it often does—develop on a profound knowledge of the parts, the parts that were preparatory in the elaboration of the whole, e.g. elements of prior theory, are not the same parts after the whole has been established. The parts that go into the creative transformation, e.g. data, tradition, technique, differ from the parts that emerge from the new concept. After the new concept is grasped, the parts are conceived differently and thus are different parts. The parts are reconfigured or transformed by the new mode of understanding.

In a discussion of scientific creativity, David Bohm has written that, "in most cases it is not (an) experiment ... that falsifies earlier theories and conceptions; rather it is some new understanding which arises in response to reflection on the *total* situation".[7] To reflect on a total situation or to achieve a new understanding is to grasp an original perspective, and this perspective is a conceptual whole with the potential to develop into the parts that it anticipates.

The intuition, grasp or total understanding surfaces to awareness as its implicit "content" becomes more fact-like, thus more explicit. The moment when the conception becomes explicit is the inspiration or total understanding. Still, even at this stage, the explicitness is not that of actual fact. The creative product is generated through a re-surfacing of fragments of the original concept.

Inspiration, then, is the presentation in the mind of a conception too replete for expression that enfolds the entirety of the work. This conception is replaced by part-concepts that, through repeated presentations, empty into the art, the theory or the science. In the translation of a concept "in the head" to a canvas or a page, anticipatory concepts develop to actual facts through the iterated analysis of conceptual wholes.

---

[7] Bohm, D. (1964) On the problem of truth and understanding in science. In: Bunge, M. (ed.) *The Critical Approach to Science and Philosophy*, Free Press, Glencoe, London.

A more precise formulation, then, is that creativity is the articulation of transformed parts out of novel wholes, where the novelty in the parts owes to the constraint of the whole that configures them, and the novelty of the whole is the originality and degree of empowerment given to the parts that emerge.[8] Emergence is the key. It is a general principle of physical and mental process that every effect is in some degree, however slight, an emergent whole,[9] and that every whole is an emergent part derived from a larger, antecedent whole. The creative process can be studied from these several aspects: concept formation as the basis of creative thinking; this process as an instance of the whole/part relation; and the whole/part relation as the basis of creativity in the mind and creative advance in the world.

II

Pathology of concepts

Disorders of thinking are important to study because pathology exposes in the symptoms of abnormal thinking the very mechanisms that are involved covertly in the normal thought process. The neurologist most closely associated with the topic of thought and its disorders was Kurt Goldstein,[10] though the conditions he studied the most thoroughly, the disturbances of language and perception, he considered to be disorders of the *instrumentalities* of thought, not disorders of thought itself. Goldstein described an impairment of the abstract attitude in certain aphasic, amnestic, frontal lobe and other cases which he believed to reflect the disruption of a thought process prior to its implementation in speech, action or perception. For Goldstein, the abstract attitude, i.e. conceptual or categorical thinking, was basic for the establishment of a voluntary mental set, for shifting from one set to another, for grasping a whole, breaking it into its parts and recombining them, and for holding multiple aspects of a situation in mind simultaneously.

The abstract/concrete dichotomy deserves closer attention for it is linked to the whole/part problem and is a central aspect of thought and its disorders. In concrete behavior, the category cannot be accessed from the instance, e.g. the color red is not abstracted independent of a red apple. If the patient can accomplish this task, he may not arrive at two or more categories from an instance, e.g. that an apple is a member of the category of shape (round), of color and of food. The impairment of abstraction can be rela-

[8] From this perspective, computer simulations of creative thought [e.g. Turner, S. (1994) *The Creative Process*, Erlbaum, New Jersey] will be of interest only to the extent they capture the natural state, i.e. whole-part shifts.

[9] Hartshorne, C. (1970) *Creative Synthesis and Philosophic Method*, SCM Press, London.

[10] Goldstein, K. (1948) *Language and Language Disturbances*, Grune and Stratton, New York; Goldstein, K. (1939–1963) *The Organism*, Beacon Press, Boston.

tively selective, as in defects of color naming or sorting, e.g. naming coloured objects or grouping them in color categories, or it can be generalized and affect a great many perceptual tasks, as in frontal lobe patients.

In the latter cases, the deficiency is the basis for impairments on tests such as the Wisconsin Card Sorting Test, in which the patient is required to sort objects along several different dimensions. Given an instance of a category, e.g. shown a round red object and asked to group it with similar objects, the patient cannot derive the target category from the member items. Nor can the patient sort according to several dimensions (color, shape, etc.). Such patients may verbalize the correct strategy for the task but not implement it in action. Luria[11] referred to this as a disorder of verbal regulation, i.e. a dissociation between thought and action.

The opposite behavior is also commonly observed. For example, on a naming or reading task an aphasic or dyslexic patient may sample a word or object category, e.g. saying grapefruit instead of apple, or reading zebra instead of horse, relying on the lexical-semantic and, less often, perceptual features of the object, or giving responses such as eat or tree for apple, where the situation or the nexus to experience is prominent. One might speculate that the correct category is realized, e.g. fruit, but the instance, e.g. apple, is not derived out of the category. In these cases, the difficulty is in going from category to exemplar, not the reverse. Indeed, there are cases where the patient is able to sort written words and objects according to their category without being able to recognize, not just name, the categorized item.[12]

The inability to go from the category to the item, i.e. having the category (fruit) but not the item (apple), suggests a priority of wholes in the progression from whole to part. However, those cases with an inability to derive the category from the item do not imply the opposite process, i.e. going from item to category, or that the patient recognizes the items but not the concepts or categories that stand behind them. Every item (word, object, act) develops out of a category or concept. Sorting tests involve object and category identification. A patient asked to group various objects into their categories must access multiple phases in the same object, i.e. recognize both the category and the item. There are cases (agnosic) who "lose" the concept of the object and are unable to identify it. The object name cannot be "found" unless the concept or category to which the object belongs has already been traversed.

Brain-damaged patients without aphasia have difficulty with polysematous words. For example, given a word such as bank, and asked to point to words such as money, river, etc., they will often select only one meaning. Other cases, asked the color of an orange, may say "yellow" or "red". A similar phenomenon occurs with verbal nouns, e.g. what do you shovel snow

[11] Luria, A.R. (1966) *Higher Cortical Functions in Man*, Basic Books, New York.

[12] Schweiger, A., Chobor, K. and Brown, J.W. (1995) From diffuse meaning to phonology. Presentation, International Neurolinguistic Society, Cracow, Sept.

with? Such patients are unable to deal with more than one meaning or interpretation at a time. There is an inability to revive the alternative concept, perhaps due to blocking or persistence of the initial interpretation out of which the polysematous item develops. The fact that patients with severe aphasia (or schizophrenia) show enhanced semantic priming[13] even for words they do not recognize or misidentify, or that one meaning of a polysematous word may prime the other, or that patients with cortical visual defects can extract the meaning of words they do not consciously perceive,[14] argues strongly for early or preliminary access to the semantic or object category—even to many related semantic categories, i.e. shades of meaning—with subsequent derivation of the specific exemplar or member item. This, I would argue, is another instance of the transition from whole to part.

Most patients recognize the instance as a feature of the category, but there are cases, such as those described by Weinstein and Kahn[15] where disparate objects are identified on the basis of shared or overlapping features, e.g. a patient who names a doctor as a butcher based on the white jacket or perhaps the comparable level of skill. This is not just facetiousness; the patient believes in the identity of the two objects.

A similar behavior in schizophrenics (the von Domarus effect)[16] is the identification of disparate topics on the basis of shared attributes. For example, an apple is thought to be poisonous due to the shared property of, say, being able to be ingested or on the basis of literary mediations. The motivation for the identification, e.g. paranoid ideation, is a sign of the depth at which the error arises. The deeper the origin of the error, as in dream or schizophrenic thought, the more it samples personal memory and subconscious cognition.[17] This is the irrationality of psychotic thought.

In hallucination or dream, shared perceptual or semantic features provide the basis for the substitution of objects, e.g. a knife may be a symbol for a penis based on the shared attribute of shape, penetration, etc. A similar mechanism is at play in so-called schizophrenic paralogic, e.g. Mary is a virgin/I am a virgin/I am the virgin Mary. A similarity by way of a common

---

[13] Increased speed of recognition of a word, e.g. nurse, by prior activation of the category through exposure of a related word, e.g. doctor.

[14] Marcel, A. (1988) Phenomenal experience and functionalism. In: Marcel, A. and Bisiach, E. (1988) *Consciousness in Contemporary Science*, Clarendon, Oxford.

[15] Weinstein, E. and Kahn, R. (1952) Nonaphasic misnaming (paraphasia) in organic brain disease. *Archives of Neurology and Psychiatry* 67:72–80.

[16] Arieti, S. (1967) *The Intrapsychic Self*, Basic Books, New York.

[17] The depth of origin accounts for other features of the schizophrenic disorder. Hallucinations are truncated object developments, delusions the play of word-meaning relations unencumbered by the drive toward denotation. Derealization is the presentiment of the cognitive origins of objects. The gaining of reality by concepts accompanies the incomplete exteriorization and "detachment" of objects. The loss of reality of objects coincides with increased affect in those concepts that are their precursors. The passive self of the dream invades the waking experience and invites the delusion that the individual is a victim for his own images to persecute.

feature or predicate suffices for an identity of otherwise different objects or subjects.

These examples, I believe, are all variations on the same theme, which can take *inter alia* the following forms:

1. Failure to go from one or more instances (items, attributes, predicates, features) to their context or background category, and the reverse, failure to recognize the item given the category.
2. Inability to stabilize or apprehend the partial relatedness of an item given its category, so that different categories (objects, subjects) are identified on the basis of common attributes
3. Failure to apprehend an item as belonging to multiple categories or to apprehend an object or category as having multiple features or meanings.
4. Inability or lability of shift from one context/item pair to another; inability to suppress one context or meaning in favor of another.
5. Ability to access the category but not identify the item
6. Ability to identify the feature but not the object, or the object but not the category.

The different manifestations of context/item or category/member relations are complex, but a thorough analysis of this topic should clarify the relation of pathological to normal thinking. What is item or member and what is context or category accounts for the difference in symptomatology across patients and from the neurological to the psychiatric series. In psychiatric disorders, shared features tend to be the basis for an identification or substitution of categories, whereas in neurological cases, a single category tends to be sampled for items having features in common with the target object. The continuum from the psychiatric to the neurological reflects the depth of cognition sampled—thus the affective intensity and/or delusional quality of the error—and the content or modality, i.e. the degree of generality or specificity to language, perception or action, which is also a function of the depth of the error.

Every item is an element for a larger domain of study or a ground for a further analysis. The transformation of context to item or ground to figure is bottomless. Moreover, the context/item or concept/feature relation, as Goldstein noted, is a relation between wholes and parts. The part-whole relation is also central to schizophrenic paralogic. Arieti noted that "the more difficult it is to abstract a part from wholes, the stronger is the tendency to identify the wholes which have that part in common". Moreover, the identification of topics in the von Domarus effect is not due to "shifting cathexes", as in psychoanalytic theory, but is based on the "cognitive equivalence of members of a primary class".[18]

I believe the whole-part transition is in a direction from whole to

[18] Arieti, S. op. cit. p. 277.

part.[19] The reverse direction, the construction or emergence of wholes from parts, is not the opposite of that from whole to part. The part-to-whole transition is the replacement of the mental state that leads to a given part by an antecedent whole that is reconfigured in the next mental state. A continuous whole-to-part shift is the process-equivalent of the laying down of the mental state. The opposite direction, the shift from part to whole, requires this process to be reversible. Since the process is linked to the asymmetry of time awareness, a shift from part-to-whole might entail a reversibility of subjective time.

Pathology unveils normal process. That is the importance of the pathological material. The varied relations of parts and wholes in the pathology of lexical and object concepts provides a basis for thinking about normal cognition[20] and creative ideation. The possibility that the part-whole relation is fundamental was first thoroughly explored by the Gestalt psychologists. Thus, Wertheimer argued that thinking concerns "the relations between parts and wholes ... involving operations as to the place, role, function of a part in its whole". Among these relations are the division of wholes into parts (subwholes), seeing the parts together without losing sight of the whole, and the achievement of closure in a "good gestalt".[21]

III

Metaphor and whole-part relations

Simile and metaphor are ubiquitous phenomena the importance of which to psychology can be appreciated by just one example, the metaphor of the brain as or *like* a machine (computer, hologram) that has been investigated as literal fact.[22] Metaphor develops out of *perceptual* part-whole relations fundamental to the human conceptual system.[23] These relations are primary. One could discuss creativity in terms of lexical concepts and still not tap the pre-lexical sources of the creative imagination. Metaphor is one way that language extends the figure-ground or feature-gestalt relations of spatial cognition.

In metaphor, a topic is assigned to a category in which the metaphoric vehicle is an instance. Take the example: *My doctor is a butcher*.[24] Attributes

---

[19] See also Krech, D. and Calvin, A. (1953) Levels of perceptual organization and cognition. *Journal of Abnormal Social Psychology* 48:394–400.

[20] Though their interpretation differs from my own, some recent studies are described in: Robertson, L. and Lamb, M. (1991) Neuropsychological contributions to theories of part/whole organisation. *Cognitive Psychology*, 23:299–330.

[21] Wertheimer, M. (1945) *Productive Thinking*, Harper and Row.

[22] As noted by Karl Pribram. The psychological literature on this topic is extensive, as in science more generally. See: Leary, D. (1990) *Metaphors in the History of Psychology*, Cambridge University Press; Brown, R.H. (1989) *A Poetic for Sociology: Toward a Logic of Discovery for the Human Sciences*, University of Chicago Press.

[23] Lakoff, G. (1987) *Women Fire, and Dangerous Things*, University of Chicago Press.

[24] This section is based on Glucksberg, S. and Keyser, B. (1990) Understanding metaphorical

of the metaphoric vehicle *butcher* serve as "connecting links" to the topic *doctor*, i.e. the two categories are related by virtue of shared attributes. In metaphor, the relation can be the basis of a creative use of language. When an aphasic calls a doctor a butcher, the relation is not metaphoric since the categories are identified. Whatever is metaphoric in the expression is inferred by the listener. The speaker is not attempting to convey an unusual meaning and is unaware of the error.

At one level, a metaphor is a comparison. In other comparisons, e.g. "a grapefruit is larger (sweeter, etc.) than an apple"—or in simile—"the sun is like an oven"—two similar or dissimilar items overlap or are related by an explicit attribute (size, heat, etc.). The attribute is not just a nexus uniting the categories to which the terms refer, or uniting terms in the same category, but can also serve as an *ad hoc* category, as when similar or dissimilar items are related or listed according to size, sweetness, etc.

When an aphasic names an apple a "grapefruit", he is sampling the category (context) of these items, i.e. fruit. The misnaming is an implicit comparison, since it relies on common (in-class) features for the identification. The background category of fruit is a whole that has the potential to actualize to parts. One could say the parts, e.g. grapefruit, apple, share features in the category that are the basis for the misnaming. Conversely, there is incomplete elicitation of the lexical items from the antecedent whole of the category.

When a schizophrenic calls himself "football", as did one of my patients, items from disparate categories are misidentified on the basis of common features, e.g. being "kicked around". In aphasia and schizophrenia, regardless of whether errors arise on features within or across categories, but particularly when experiential or functional attributes are involved, patients are not fully aware an error has occurred; i.e. the patient does not apprehend the relation implicit or collapsed in the error. In normal thought or language, in comparison, simile or metaphor, the individual is conscious of these relations as a device for the communication of new meaning. The inability to bring the whole-part relation to awareness is due, partly, to the inability to retain two items simultaneous with their relatedness. Certainly, the exposure and vulnerability of these relations reflects the difficulty in comprehending wholes and parts as distinct but related contents.

The apprehension of new meanings and the awareness of metaphoric (or other part-whole) relations are important differences between the pathological and the creative. Psychotic speech and certain types of aphasic jargon may resemble some forms of poetry. For example, a patient of mine with semantic jargon described his difficulty with vision as "My wires don't hire right". Another aphasic wrote that she: "found Brooklyn about her troubles, a small nature in the pink yellow garbage from motion". The schizophasic of

comparisons: beyond similarity. *Psychological Review* 97:3–18; and Glucksberg, S. (1991) Beyond literal meanings: the psychology of allusion. *Psychological Science* 2 (3) 146–152.

Arieti with word-salad said "The cow burnt the house horrend(end)ously alway". Such errors can be analysed in terms of contextual effects on word substitution.[25]

The similarities between semantic deviance in aphasia and schizophrenia with some forms of poetry can be striking, e.g. Dylan Thomas' "If my head hurts a hair's foot/ Pack back the downed bone". A more subtle anomaly of word meaning is found in verse by the schizophrenic poet Ezra Pound, e.g. "Shines in the mind of heaven/ God who made it/ More than the sun in our eye". In this passage, the object referred to is not given in the neighboring text.[26] Derailment in word meaning and the sampling of the contextual background of target lexical items can be turned to artistic effect. We assume that such writing is deliberate in the poet and involuntary in the psychotic or brain-damaged patient. However, the poet may be no less passive than the schizophrenic to the depth of origin of his or her verbal imagery but differs from the schizophrenic in being able to edit the material later on. Conversely, the aphasic or psychotic has little or no awareness of the defectiveness of the utterance and cannot play with creative output. The ability to edit is essential. Dali alludes to this in his witty comment that he was critical of his own paranoia.

Metaphor and concept formation

Ordinary conversation is often a description or recounting of one's states, opinions or experiences where the speaker is the topic, e.g. "I (remember, anticipate, enjoyed, etc.) drinking Pastis with Pierre". When the speaker detaches from the topic, and facts or events are not simply recounted, novel entities may occur.[27] For example, a relation of simile could introduce a new meaning to extrapersonal terms, e.g. "Pierre is (like) a saint". Here, Pierre shares features or attributes of saintliness, or an attribute that is a part of the whole concept of Pierre is an attribute of the whole concept of a saint. The common part brings two wholes together. In synechdoche, e.g. "The saint has departed", an entity (Pierre) is replaced by one of its features or parts, i.e. saintliness. The correlate of this in pathological states might be the identification of an object by a perceptual or functional feature, e.g. naming an apple "red" or "eat". Similarly, the replacement of one word for another in

---

[25] Aphasic neology is usually semantic *plus* phonological errors; e.g. "twas brillig and the slithy toves". As with errors word meaning, phonological errors can be interpreted in terms blends and contextual effects on phoneme production [Buckingham, H. (1994) Presentation. New York Academy Sciences].

[26] For other examples, see Critchley, M. (1967) The neurology of psychotic speech, *British Journal of Psychiatry* 110:353–364.

[27] Conceptually original statements, not just novel sentences, which are the rule in language use [Pind, J. (1994) Computational creativity: what place for literature. *Behavioral and Brain Sciences* 17:547–548].

metonymy is a common error in aphasia.

An additional step takes the expression further, e.g. "Pierre has piety without religion". This step complicates the part-whole relation in an interesting way. Now the feature *piety* that links Pierre to a saint is cleaved from another feature of saintliness, that of religiousity. The result is an expression close to irony. When the cleavage is made explicit, e.g. "Pierre is no saint", the feature, say piety or celibacy, becomes the implicit topic in a remark that turns critical. Such examples show how the play of concept-item or category-attribute relations can become very complex with *ad hoc* concepts or categories continuously being formed.[28] Of interest are studies in aphasics showing *ad hoc* categories on sorting tasks when abstract categories are unavailable, for example, sorting tiger with crocodile instead of cat based on their common ferocity.

In the example, "Pierre has piety without religion", a microgenetic account might hold that the categories of Pierre and religion parse to a featural element, *piety*, which is affirmed in the proximate section of the sentence and negated in the distal section. The fractionation of categories or wholes to their exemplars, e.g. Pierre, religion, supplies the main terms, while the shared features take on opposing contrasts. What is a whole and what is a part is arbitrary. Features can become concepts and vice versa. The *feature* piety can become the *concept* "piety" or the *category* "pious things" to which other wholes, e.g. Pierre, religion, can then relate as positive or negative features.

In bisociation,[29] a new concept arises, perhaps as an act of inspiration, in the fusion of previously unassociated concepts. While conceptual growth is probably more often the result of a gradual transformation than a sudden insight, the account of bisociation resembles that of the blending of items or categories in simile and metaphor. Whether the result is a pre-lexical concept, as in creative thought and imagery, or a more restricted phenomenon such as a new metaphor, depends on which categories, concepts or words serve as topics or vehicles, and which feature or set of features is the axis of the transposition. The profoundly creative involves concepts of breadth and/or universality in relation to the elements those concepts enclose. Schopenhauer wrote that the fundamental characteristic of genius is "always to see the universal in the particular".[30] This aspect of genius, along with the depth of creative insight (see below), is embodied in the notion that genius explores underlying concepts (intuits) while talent works with surface elements (analyses).[31]

---

[28] Barsalou, L. (1987) The instability of graded structure. In: Neisser, U. (ed.) *Concepts and conceptual development: ecological and intellectual factors in categorization*, Cambridge University Press, Cambridge.

[29] Koestler, A. (1964) *The Act of Creation*, Basic Books, New York.

[30] Schopenhauer, F. (1907–1909) *The World as Will and Idea*. Vol. 3. In R. Haldane and J. Kemp, (eds), London.

[31] See Hirsch, N. (1931) *Genius and Creative Intelliqence*, Sci-Art Publishers, Cambridge, Mass.

Categories and concepts

Part-whole relations figure in the aquisition of basic level categories, e.g. dog, car. Such categories depend on gestalt mechanisms of perceptual similarity, especially shape, while superordinate categories, e.g. animals, vehicles, tend to share functional features. Perceptual wholes and features are more salient at the basic level, e.g. the shape of a car, the wheels, engine, etc. Moreover, in such objects, "the wholes seem to be psychologically more basic than the parts".[32] Basic objects are the first to be learned in childhood. The whole-to-part transition is characteristic of early cognitive development and a part-whole relation is critical in the learning of basic objects. Part-whole relations are easier than class inclusions, and there is a shift in childhood learning from a reliance on categorical meaning to a reliance on features.[33]

For Tversky and Hemenway, the decomposition of wholes into parts is the basis on which structure is used to "link the world of appearance to the realm of action", and to comprehend, infer and predict function.[34] This relation characterizes intuition and naive induction and is the basis of novel concepts in scientific and other forms of creative thought. These authors give as an example the shift from holistic concepts of brain function to the concept of functional localization. For Lakoff, basic level objects and their parts engage action and the body schema to generate metaphor and complex categorizations.[35]

A category tends to be the more primordial entity, a concept is a more specific or individuated category. A category is a group of like things that resemble each other along some dimension, e.g. the shared features of *dogs* or *chairs*. A concept incorporates the dimension along which such resemblances are established. We say, the *concept* of a chair, i.e. the shape, the features and functions that determine what a chair is, and the *category* of chairs, i.e. the grouping of chair-like objects, or objects that satisfy the *concept* chair. The category of dogs and chairs includes by implication the perceptual features of those objects.

Animals have primitive object concepts for shapes and features. A dog can recognize the categories of dogs and chairs, prey, shelter, without a concept of the meaning of these objects other than the responses they call forth. An object concept is a whole-part relation that, when generalized over similar

---

[32] Lakoff, op. cit.

[33] See Carey, S. and Gelman, R. (1991) *The Epigenesis of Mind: Essays on Biology and Cognition*, Erlbaum, New Jersey; Markman, E. (1981) Two different principles of conceptual organization. In: M. Lamb and A. Brown (eds) *Advances in Developmental Psychology*, Erlbaum, New Jersey, 199–236. Keil, F. (1987) Conceptual development and category structure. In: Neisser, U. op. cit.

[34] Tversky, B. and Hemenway, K. (1984) Objects, parts, and categories. *Journal of Experimental Psychology: General* 113 (2) 169–193.

[35] Lakoff, G. (1987) op. cit.

objects, is the category of that object type. The configural aspect of these objects and their part-whole relations determine the object concept of dog or chair. A lexical concept, e.g. the word "dog", is also a whole-part relation. The word has the potential for different meanings and denotations. The shift from the potential of the word to a specific instantiation is a species of the whole-to-part transition.

Depth and surface

Though itself a metaphor, depth of process is important.[36] The depth of creative thought introduces a *micro*temporal, cladistic[37] or genetic dimension to category formation. The relatedness between concepts reflects their immediate prehistory, not actualities at the surface which are mere outcomes. Concepts arise in the subconscious of long-term memory organized around experiential and affective cores and traverse the dreamwork, images, symbolic and metaphoric relations, and the like, on the way to propositions and the rational or logical structures they instantiate as "facts" in the mind or the world.

Jung wrote of a "visionary mode" with its own subconscious autonomous form. In this mode, the individual is passive to the emerging creative product. The passivity is a clue to the depth of origin of the creative idea. As in hallucination and dream, the self is passive to its own emerging content. Fantasy and reverie are associated with a receptive attitude. Goethe said, "Thinking doesn't help thought"; and Beethoven, "You ask me where I get my ideas. That I cannot tell you with certainty; they come unsummoned".

I would align myself with Kris,[38] that creativity is a flight from deliberation in the service of novel concepts. The withdrawal from objects to their anticipatory constructs in spatial and imaginal thought allows a more generic concept, i.e. one with the potential to develop into different modalities, to be realized in a specific cognitive domain. This ability to *dip* into the "pool of the creative unconscious" is an uncommon experience for the average person, those who, as Wordsworth wrote, have a "mind intoxicate with present objects and the busy dance of things that pass away". The creative personality reclaims the conceptual and symbolic sources of those objects. The depth of creative work implies an engagement of fundamental aspects of the personality, whether in the sciences or in the arts. When Piaget pointed out to Einstein that his concept of spacetime resembled time perception in small children, i.e. young but not older children perceive time in terms of

---

[36] Smith, G. and Carlsson, I. (1989) *The Creative Process*, Psychological Issues 57, International Universities Press.

[37] Cladistic categorization is based on a shared derivation in contrast to categories based on overall similarity (Meyr, E. (1982) *The Growth of Biological Thought*, Belknap Press, Cambridge, 226–233; also, Lakoff op. cit., 118–121).

[38] Kris, E. (1952) *Psychoanalytic Explorations in Art*, New York, International Universities Press.

spatial relations, Einstein is reported to have wondered whether this might have been the result of his slow maturation.

IV

Creation and nature

We gain a better understanding of mental process by a study of its pathology and correlated brain mechanisms. In so doing, we move closer to a depiction of the physical, to which the mental is our only contact. If whole-part transitions are the basis of human thought and creativity, and if mind is part of the physical world, the relations that characterize mental process would be the same relations that hold for creative advance in nature.

The world of thought and perception emerges through a graded analysis of wholes into parts. In this process, an object is the outcome of change in a passage to greater definiteness. Every entity in the world is a momentary novelty. The world is never twice the same. Creativity depends on the potential of this activity to regenerate the world and the incompleteness of actualizations through which novel worlds are generated.

Science is the study of the relations of physical succession. Microgenesis is a theory of emergent recurrence. The whole-part or "many-one" relation—the elicitation of items out of contexts—is a fundamental property of mental process. Is the whole-part relation *the* underlying principle of change in mind and nature? Whitehead thought so, and wrote that creativity is "the ultimate principle by which the many, which are the universe disjunctively, become the one actual occasion, which is the universe conjunctively".[39]

Mind gives duration to the transient events it records, like the tidal waters of a rushing stream, a moment of persistence in the flux of actual events. The link from duration to creativity was the theme of Bergson's great work. He wrote, "the more we study the nature of time, the more we shall comprehend that duration means invention, the creation of forms, the continual elaboration of the absolutely new".[40] Duration is the basis of categorization and basic categories are the nuclei of primitive concepts that give rise to objects. An object is first a concept in memory before it is an object in the world. The recognition of the object is by way of the concept of that object that summons the object up. Memory does not hold on to the world but creates it and it is memory, or the process that makes memory possible, that sustains the world over its momentary instantiations. In this way, through the continual formation of new concepts, mind gives meaning and stability to the raw succession of physical states. This is the expression of creativity in physi-

---

[39] Whitehead, A.N. (1929) *Process and Reality* p. 31; Macmillan, New York. Also: Pols, E. (1967) *Whitehead's Metaphysics*, Southern Illinois University Press, Carbondale.

[40] Bergson, H. op. cit.

cal process. After all is said, the creative life is the potential of concepts to expand the novelty of physical succession to the generation of abstract entities that endure.

New York University Medical Center, USA

DONALD G. SAARI

A FOURTH GRADE EXPERIENCE

"We've been trying to tell you you're wrong!" With this reprimand, a nine-year old girl, hand on hip, expressed the class' exasperation with the slow-witted mathematics professor who couldn't understand the obvious. That morning I got a lesson about whether the creative-like behavior of children reflects creativity or an inability to discriminate.

Dr. Diane Briars, Director of Mathematics, invited me to discuss modern mathematics with several student groups as part of the Pittsburgh Public Schools' observance of the 1991 national "Mathematics Awareness Week". It was easy to prepare for the older students—a grab bag of unusual examples from "chaos", mathematical astronomy, and statistics could illustrate mathematical themes while capturing their imagination and encouraging creative responses. But I fought to disguise my growing panic when informed that I was to talk to a fourth grade class. *What does one say to nine-year olds?*

The mathematics of voting serves as a useful vehicle to introduce mathematical concepts while stimulating creative reactions. So, to survive my allotted forty minutes, I described a counting problem involving an hypothetical group of fifteen children permitted to watch only one TV show for the evening. Of these children,

	Best	Second best	Last
6 preferred	Alf	Flash	Bill Cosby
5 preferred	Bill Cosby	Flash	Alf
4 preferred	Flash	Bill Cosby	Alf

Which show should they watch?

For us, the answer obviously is "Alf" as dictated by the election outcome

Alf is preferred to Bill Cosby is preferred to Flash

with the 6:5:4 vote. My worse fears were realized when immediately after posing the problem the class answered with that droning sound of a drill of addition facts, "*Flash*". Were these students too young to understand the simple relationship between counting and voting? Were they incapable of

My visit to Pittsburgh was supported, in part, by an NSF grant to the University of Pittsburgh. The research results described in this article were supported by an NSF Grant and by my Arthur and Gladys Pancoe Professorship in Mathematics.

Å. E. Andersson and N.-E. Sahlin (eds.), The Complexity of Creativity, 51–58.

handling abstract issues that differed from their daily routine? Why should we expect useful innovative responses from young children?

Calling on any poise hopefully gained through years of teaching, I challenged their conclusion. Their instant response, "Well, you see, to choose the best show you have to count the number of times each show is in first place and how many times the kids like it next best". "If you look, some kids like Flash the best and all other kids like Flash next best, but all other shows many kids like the worst". Maybe they did understand. At least their answers used the data rather than just voicing personal favorites.

I persuaded them to indulge in my suggestion ("Let's count how many people like each show the best!") of using the standard election procedure. While tending to treat my suggestion as a silly aside, they politely agreed that it would define the ranking

Alf is preferred to Bill Cosby is preferred to Flash.

Then, to complicate the story, I announced that "Alf is canceled tonight. Now what show should these kids watch?" Again, this is trivial for us; the obvious choice is the second place "Bill Cosby". But their immediate response, with a frightening agreement, was "Flash!"

After challenging this second answer, even more children entered the debate by countering "You just can't count who likes what show the best, you have to see what they like next best too". "Your counting way that makes Alf best and Bill Cosby next best is silly". "Count! If you count you'll see more kids like Flash than Bill Cosby". They are correct. In this example the last-place Flash defeats second-place Bill Cosby by the surprisingly large vote of 10 to 5. Before I could use my intended punch lines that "last place Flash is preferred even to first place Alf" and "more people prefer second place Bill Cosby to first place Alf", most of the children proved they already had completed the analysis by triumphantly calling out "Look, count! See, they like

Flash best, Bill Cosby next, and Alf last!"

"We tried to tell you, these kids like 'Flash' best!" It was when I stood silently astonished by their insight and quickness, rather than the intended conclusions of my concocted example, that I got my reprimand. More were to follow.

VOTING

This example identifies a serious flaw of our standard tool of democracy. As suggested by the story, it is not difficult to find actual elections where the plurality winner is the candidate the voters actually view as being inferior

—while the candidate the voters view as superior ends up losing. To find an example, just check closely contested elections with three or more candidates; it will not be long before you find one. But, this flaw cannot reside in the voters' preferences, so it must be an artifact of the voting method. What should be done?

This example defines an important, easily introduced but real conflict. Because the source of the problem, leave alone a resolution, is not obvious, it offers a way to examine "creativity". Moreover, this is an important problem; the search for "better" election procedures has been a serious political concern throughout the centuries. (See [Chap. 1, S].) For example, almost a millennium ago the process of choosing the simple majority winner caused strife and even violence within the Catholic Church leading to competing claims of who is the "real Pope". These conflicts ceased after the design and adoption in 1179 of the selection process still in use. (To be elected Pope you need one more than ⅔ of the votes cast by the Cardinals.)

The modern study of elections was initiated by the French mathematician J.C. Borda in his presentation to the French Academy of Science in 1770. He, along with fellow academy members such as Condorcet and Laplace, recognized that while it appears to be simple to choose an election procedure, it is not; it is mathematically complicated. Indeed, since the 1780's, this topic has received the attention of hundreds of researchers writing thousands upon thousands of published papers, yet much of the mystery remains untouched.

RESOLUTION

Presumably, the reader is interested in creativity. If so, before reading any further, you are invited to find a "correct", or at least a "better" voting procedure.

Actually, this challenge is unfair; the complexity of the problem causes even professionals to shy away from offering proposals. So, it was more out of curiosity than with any expectation of a response that I posed it to my fourth-grade subjects. "How should these children choose the TV show to watch?" Quietly but surely Cara answered, "Well, I think we should give 3 points to the show we like the very best and 2 points to the show we like next and only 1 point to the show we like the last. This way we can also tell what other shows the kids like other than their best one". This is Borda's procedure; a process that only recently has been shown [S] to be optimal for many reasons. Abe, an obviously bright boy who knew it, countered with "I want to give 1 point to our best show and 0 points to our next show and –1 points to our last show". A nine-year old advocating a procedure based on a *negative* number of points! Then Susan suggested, "How about giving 2 points to our best show and 1 point to the next show and no points for the last show". By now everyone wanted their say, but several children cut off further discussion by pointing out that each of these procedures yield the same election outcome.

They were, again, correct. As these children argued, it doesn't matter how many points are assigned to each candidate; the critical factor is the point spread between the first and second ranked candidates, and the second and third ranked candidates. While their wording reflected their tender age, their surprisingly sophisticated argument emphasized that the point differential is what really determines an election outcome! Because the differential for each of the three proposed methods is a single point, the election rankings must remain the same. In fact, by using any of the Borda – Cara – Abe – Susan procedures with the above example the election ranking is

Flash in first place, Bill Cosby is in second place, and Alf in last

—an election ranking that reverses the plurality outcome and is totally consistent with the pairwise rankings of the TV shows. Using Cara's 3—2—1 method, the tally sheet is

Voters	Alf	Bill Cosby	Flash
6	18	6	12
5	5	15	10
4	4	8	12
Total	27	29	34

Surprisingly, only Borda's method [S] exhibits this consistency. With any other way to tally the ballots where the differential between points is not a constant, no relationship need exist among the election rankings of the candidates and the pairs of candidates! Instead, choose rankings for each pair of candidates in an arbitrary fashion (say, by flipping a coin) and choose any ranking for the three candidates (say, by consulting a lotto outcome). The disturbing fact is that examples exist (maybe requiring more than 15 voters) where each voter has a specified ranking of the candidates and where the actual election ranking for each set of candidates is the arbitrarily selected one! This does not inspire confidence in the outcomes of these widely used instruments of democracy. Only Borda's method is immune from all possible kinds of electoral chaos.

It is of importance to this discussion about creativity that these children understood the basic reason for the success of Borda's method. I discovered this by challenging their choice of 3, 2, 1 by suggesting we should use some other choice such as 6, 1, 0. The children rebelled; they argued that this choice provides too little weight to the second place candidate. They argued that fairness required the symmetry provided by Borda's weights. Indeed, this symmetry is the mathematical source of the consistency of Borda's method.

While the argument is technically difficult, these children sensed it; I could not determine how.

CONDORCET'S EXAMPLE

I still had twenty minutes left! So, I started constructing a version of an example advanced in the 1780's by Condorcet; a puzzling example that continues to motivate the voting and decision literature. After writing down

	Best	Second best	Last
5 preferred	Alf	Bill Cosby	Flash
5 preferred	Bill Cosby	Flash	Alf

a girl in the front volunteered, "And you are going to say next that 5 like Flash best, Alf next, and Bill Cosby last". She correctly recognized the pattern. Can you? (A test is if you can construct a four candidate example.) A version of the Condorcet example has

	Best	Second best	Last
5 preferred	Alf	Bill Cosby	Flash
5 preferred	Bill Cosby	Flash	Alf
5 preferred	Flash	Alf	Bill Cosby

A simple count shows that by an overwhelming 10 to 5 vote these people prefer Alf to Bill Cosby and by the same 10 to 5 margin they like Bill Cosby over Flash. Presumably this means they prefer Alf to Flash, so Alf is their top-choice. Yet, when asked which show these kids prefer, the class answered, "Nobody is best; they are all the same". A patient student, in a patronizing fashion, summarized the class' arguments by slowly explaining, "See. Each show is the same number of times in top place and in second place and in last place. That is why there is no favorite; they're all the same".

Sticking to my guns, I insisted on comparing Alf with Bill Cosby to demonstrate the 10 to 5 conclusion; an assertion greeted by an avalanche of outbursts, "Yes, and Bill Cosby will be better than Flash and Flash will be better than Alf by the same numbers". They are correct; Condorcet's example illustrates that cycles—even of landslide proportions—can arise when we retreat to our comfortable procedure of the majority vote of pairs. One incredibly small boy took pity on me by carefully offering, "Let me explain. Nobody is better; they are all the same. It's like the rock and the sizzors and the paper. The rock can dull the sizzors and the sizzors can cut the paper and the paper can cover the rock, so nothing is better than the others". He was correct, of course. They are the same.

To fill the remaining minutes, I offered a dollar reward for each solution of a (recently solved) research problem, and was relieved when my wallet was saved by the class bell.

CREATIVITY

It has often been noted how the unusual, surprising comments of children resemble creativity. But, there are protests; while many responses of children are surprisingly innovative, are they really creative?[1] To test, we could apply standards. For instance, if we accept Vernon's ([V, p. 94]) "creative product" definition where "[c]reativity means a person's capability to produce new or original ideas, insights, restructuring, inventions, or artistic objects, which are accepted by experts as being of scientific, aesthetic, social, or technological value", then children usually fail because most of them are not sufficiently sophisticated to match this measure of utility.

Alternatively, we could emphasize the "creative process", Here, it is easy to equate the standard creativity enhancing procedures—such as divergent thinking, brainstorming, the acceptance of novelty—with child-like behavior and their partial suspension of evaluation. But, creativity requires something to be accomplished, so higher standards normally are applied to creative thinking. To be judged creative, it does not suffice for a new idea or concept to be novel; it must be distinctly different from what the person previously knew. Rather than providing a step beyond what was known, creativity requires a leap. This creates a difficulty. With the typical types of problems proposed to children (such as finding different uses for a tin can, or the number of rabbits in a picture), how do we know what the children already know or whether they care?

My accident of using the voting problem may offer a partial resolution for these difficulties. Not only are these issues unknown to young children, but they have confused generations of professionals from mathematics, economics, political science, and other research areas. So, the arguments and resolutions proposed by these nine-year olds satisfy both the product and process criteria for creativity. What surprised me is that while these concerns are not in anyone's store of knowledge, they did not stymie the fourth graders! With innocence coupled with awakened curiosity, with insight freed from our myopic view blurred by years of blind acceptance of a standard but flawed election procedure, these fourth graders cut through the conceptual difficulties to achieve critical understanding.

I suspect that the pressures of the playground and group interactions formed a sense of fairness within these children, but a sense that had yet to be codified. It was this well of undeveloped understanding that I accidentally

---

[1] Indeed, at the 1994 *Cognition and Creativity Workshop* in Venice, Italy, this topic was a point of contention.

tapped into with my example. Probably because these children did not have a background in this area, they could not rely upon a mechanical conclusion (as we often do). Instead, they needed to search and develop answers by examining the data. Were these children creative? Absolutely!

This experience suggests using voting examples to examine "creativity". To explore this possibility, I tried, with varying levels of success, similar examples in other classrooms. When I dealt with young children, from around seven to nine or ten, the results were somewhat similar to my Pittsburgh experience. But once the children are, say, teenagers, they seem to be sufficiently wedded to the standard voting procedure that they have difficulties seeing through the issues.

I also discovered that a critical factor is the manner in which the class is conducted. When the class is run in a fashion to encourage "creativity enhancing procedures", then creative responses are forthcoming. In particular, a useful approach was to create the image that I was "baffled" by the issues; the young students really wanted to help me. My speculation is that by being "confused", the students did not view me as being an all knowing authority figure with the "correct" answer; they lost their inhibition. Because they did not need to worry whether they were "wrong", imaginative answers were forthcoming. It became socially acceptable to become risk takers, to make errs.

A couple of my mistakes in running sessions reinforces this suspicion. One was when progress in a class was abruptly disrupted after the teacher reprimanded a child who was mildly fooling around. Instantly, her action promoted her to "the authority figure" on the scene. No longer was it socially acceptable to be wrong because the students now looked to her whenever responding; their suggestions became cautious—and boring. Another time, I erred by approving rather than cautiously wondering about their invention of Borda's method. (Surprisingly, while other suggestions were made, Borda's method always was a proposal!) My certifying comments "blew my cover" because this statement identified me as an authority figure who accepted this method. This seemed to kill their incentive to develop creative arguments to prove that their choice was better than, say, the weights 6, 1, 0. The tone switched; pensive arguments were replaced with platitudes. Even worse, instead of thoughtfully trying to help me, some adopted a "shotgun" strategy where they would quickly propose many answers expecting me to select the correct one.

On the other hand, one class was so excited by their discoveries about the flaws of voting that they wanted to write a class letter to the US Congress to explain what Congress and the country was doing wrong! (I do not know if they did.) Also, I was pleasantly surprised that in each classroom there always was at least one student who worried whether the proposed procedures were specific to a particular choice of preferences rather than resolving a general problem. (By reading the voting literature, with its dependency on specific examples, I cannot say with comfort that this subtle point is appreciated by all of the professionals in this area.) Always some student worried whether their proposed procedures would fail with other choices of preferences. To test this, they tried to construct new examples. (But, their standard examples did not succeed in exposing the difficulties.)

There is another issue here. If the response of these classes indicate what is possible at such a young age, then we must wonder what it is we do to inhibit the natural creativity and inventiveness exhibited by these children. We must wonder how our usual classroom approach of imposing solutions through authority rather than exploring ideas to generate and understand answers can lead us to mediocrity. Children can be bright, but by emphasizing rote and accepted procedures, rather than looking fresh at the data, we may be encouraging them to adopt a strategy which discourages "creativity". Ending with an unimaginative but appropriate cliche, we must wonder what can we do to be part of the solution, rather than causing the problem.

Department of Mathematics
Northwestern University, USA

REFERENCES

[S] Saari, D.G., *Basic Geometry of Voting*, Springer-Verlag, 1995.

[V] Vernon, P. "The nature-nurture problem in creativity" *Handbook of Creativity*, ed. C. Reynolds, Plenum (New York), 1989, 93–109.

N.-E. SAHLIN

VALUE-CHANGE AND CREATIVITY

England, during the last decades of the 19th century, is a society full of fixed expectations and rigid values. The Victorian way of thinking was, for example, permeated by the idea of an objective sexual norm. Theresa Berkley, a well-known businesswoman of the time, knew how to make money out of prejudices. At her establishment the frustrated Victorian gentleman could choose from an ample menu of sexual recreation. Her brothels for flagellants offered an impressive repertoire of services: one could get "birched, whipped, fustigated, scourged, needle-picked, half-hung, holly-brushed, furse-brushed, butcher-brushed, stinging-nettled, curry-combed, phlebotomized". But it doesn't matter how good you are, you do not get rich flogging them one at a time. Theresa Berkley realized this and invented history's first flogging machine, the Berkley Horse, and made her self a small fortune.[1]

Is not Theresa Berkley an example of a truly creative mind? The answer is no doubt Yes. It is not what she accomplished that is important, it is how she did it. Compare, for example, Berkley with her customers, these esteemed citizens of the Victorian society, prisoners of their own rigid system of norms and values. What she did, but they did not, was to break through the norms and values. A probable explanation of Theresa Berkley's success is that she came to entertain a different system of expectations, a basis that allowed her to make what to us looks like a creative move.

The development of mathematics is another example of how the change of representation yields new and fascinating results. Around the turn of the 19th century quite a remarkable number of interesting mathematical theorems were proved. Vague ideas and conjectures were transformed into a fertile formal language and were finally solved. The new language gave the mathematician a new way to formulate the means as well as new and highly efficient methods to prove his theorems.

In his doctoral thesis (1799) the young Carl Friedrich Gauss proved that for any algebraic equation, there exists at least one root (i.e. a complex number such that $f(r) = 0$). This theorem allows us to prove what is known as the fundamental theorem of algebra, i.e. that every polynomial of degree n can be factored into the product of exactly n factors, the roots of the

[1] See Englund (1991) and Pearsall (1993).

Å. E. Andersson and N.-E. Sahlin (eds.), The Complexity of Creativity, 59–66.

equation. This is an existence theorem. It tells us what there is. But it is definitely not a constructive theorem, in the sense that it tells us how to find the solution; all it tells us is that something exists.

By an intricate and almost non-formal way of reasoning mathematicians had as early as in the sixteenth century come to see that algebraic equations up to the fourth degree can be solved by a sequence of operations; roots are extracted and simple standard algebraic operations are employed. The complete series of moves in this operation gives us a solution by radicals. Today it is well-known that great mathematicians squandered considerable time and energy trying to extend this method to algebraic equations of degree 5 and higher. History thus reveals that mathematicians obviously expected a solution to the problem. For them it was just a matter of time and energy. Today we know that they were prisoners of the classic-mathematical system of expectations and values.

Niels Abel, however, came to entertain a rather different system of expectations. At one point he thought he had a solution to the problem, but the proof did not hold water. After that, unlike his colleagues, he did not expect a traditional constructive solution to be found. What he instead anticipated was that there might well be no solution by means of rational operations and radicals. In fact he proved that there is no such solution.

What then made this important theorem possible. Gauss entered the field of mathematics after what might be called the heyday of calculation. A rather full-fledged formalism had finally replaced simple arithmetical computing. The young Gauss walked into a new world of mathematics. What had emerged was an entirely new representation of the subject, providing new tools and methods for proving theorems. A Cardano or a Tartaglia had no such tools or methods; they had to trust their arithmetical cleverness. They solved equations of the 3rd and 4th degree by highly complicated formulas, but still variations on a well-known theme, i.e. how to solve the quadratic equation. The ability to represent mathematical problems, not by numbers but variables, is an enormous breakthrough in the history of mathematics. The mathematician is for the first time given an apparatus by which he or she can formulate problems in a very precise way. He or she can more easily discern new patterns, make generalizations and abstractions, and thus come up with bold conjectures. Furthermore, the new foundation gave them the tools and methods to prove their conjectures.

Concerning creativity, Abel's impossibility theorem is far more interesting than Gauss's theorem. Gauss was given a new representation by a community of creative mathematicians. He was given virgin soil that he skilfully cultivated and harvested. But Abel saw that however good a farmer Gauss was there are certain things that cannot be harvested, at least not with the tools he had. It is not a question of trying harder, spending more time on the problem — it simply cannot be done. What you want to accomplish is out of reach. Abel tells us something about the confines of knowledge.

This is a revolutionary result. In the garden of mathematics there is an apple-tree, full of beautiful red apples. What Abel teaches us is that some of these apples cannot be reached with our traditional tools and methods. We might know that some of them cannot be picked because it is simply too bushy where they are, but we can lop the tree. In other cases we cannot reach them because the ladders we have are not high enough, but we can put them one on top of the other. And others still because they simply are not on the tree. Abel's argument, however, is more profound. He told Gauss and us that regardless of how skilled and creative we are, relying on our standard tools and methods, there are quite a few apples on this particular tree that we will never be able to pick. Occasionally we can even see them and smell them, maybe even touch them with the tips of our fingers, but we will never be capable of picking them. In fact, on this tree, there are only four apples within our reach. (It makes sense to compare Abel's result with Gödel's well-known theorem. Gödel taught us that there are theorems of mathematics that are true, but not provable.)

Abel (in this particular area of mathematics, I should add), but not Gauss, taught the mathematician that there are things that cannot be done. He turned the expectations of mathematicians outside in. He gave them a totally new representation of their subject. We understand the importance of Abel's work if we remember that the type of question he worked on is the hub of modern algebra and group theory. A few years after Abel proved his theorem, several classical problems were "solved", or rather shown not to have a solution; for example, the problems of doubling the cube and trisecting the angle.[2]

'How much of the mind is a computer?', asks D.H. Mellor; and his answer can be summarized: 'but a fraction'. Mellor's arguments are important in discussing the question of creativity.[3]

Propositions represent facts. It is facts that make propositions true or false. It also makes sense to take propositions to be the content of our propositional attitudes; e.g. the content of our beliefs and desires.

To believe something, is to take it for true. Believing in *p* does represent the proposition as true. However, to desire that *p*, does not represent the proposition as true, it does not take *p* for a fact. Or, as Mellor puts it, 'to take *p* for a fact just is to believe it, whether or not one has any desire, hope, fear or any other attitude towards it'.[4]

G.E. Moore taught us that it does not make sense to say that '*p* is true, but I do not believe it'. Imagine, for example, Abel saying that,

> it is true that algebraic equations of degree 5 and higher can be solved by a variation of the traditional sequence of operations used to solve equations of degree less than 5, but I do not believe it.

---

[2] See, for example, Courant and Robbins (1969).

[3] Mellor (1991).

[4] Page 79.

It is indeed an absurd statement. However, there is nothing absurd in saying that,

it is true that algebraic equations of degree 5 and higher can be solved by a variation of the traditional sequence of operations used to solve equations of degree less than 5, but I do not desire (like, fear, etc.) it.

Similarly, there is nothing wrong about being afraid that p & q, but not being afraid that p, and not afraid that q. But it is absurd to believe that p & q, but not believe that p. If I take the conjunction to be a fact, I must believe in the conjuncts. Thus, as Mellor points out, Moore's so-called 'paradox of belief' simply has no analogue for any other propositional attitude.

Our beliefs aim for truth. Furthermore, making inferences, processing our beliefs, also aims for truth; simply because what they deliver is belief. Information processing, or computing, is truth tracking. It is like taking the train from London to Cambridge. The first time you do it you might find it exciting. You see things you have not seen before. But it leaves little room for creativity. Once you are on the train the trip is laid out. You have to follow the tracks. You cannot say, let's turn West at Audley End (Saffron Walden).

Mathematics and logic are full of this type of humdrum journeys. Reading *Principia Mathematica*'s three volumes of proofs for the first time might impress you. However, it should not take more than a few proofs to realize that you are following the tracks of propositional calculus; once you are on the track there is not an ounce of creativity. Similarly, almost all Henkin-type completeness theorems of logic, are but simple track-following.

But, as Mellor teaches us, the psychology of attitudes other than belief is not computational (in any serious sense); they are not at all like a boring journey from London to Cambridge, simply because they have very little to do with truth-tracking.

It is thus obvious that what has been called computational psychology, the study of the computational processes whereby mental representations are formed and transformed, never will disclose the enigma of creativity. The value aspects are too important.

But, one might argue, this criticism is valid only if we are working within the framework of classical logic. What about nonmonotonic reasoning, isn't this type of inference a possible sign of creativity? From our perspective a person not accepting what the paradox of belief tells us, might well be looked upon as making nonmonotonic inferences.[5]

Let '**G***x*' stand for '*x* is a good mathematician', '**M***x*' stand for '*x* has the

[5] Following Gärdenfors (1993). A nonmonotonic inference is one that can be overthrown by new information. What characterizes a monotonic inference is that if B follows from A, then B follows from A and C, for any C. But in the case of nonmonotonic inferences B might not follow from A and C.

methods necessary to find the radicals', and '**S***x*' stand for '*x* solves the equation of degree 5'. Let us also assume that *A* expects it to be the case that: 1) if **G***m*, then it is not the case that **S***m* (i.e. if *m* is a good mathematician, then he will not find a solution); and 2) if **G***m* and **M***m*, then **S***m* (i.e. if *m* is a good mathematician and has the methods, then *m* solves the equation). If we make the additional assumption that *A*'s set of beliefs is closed under logical consequences, this set will also contain the expectation that 'if **G***m*, then not-**M***m*' (i.e. if *m* is a good mathematician, then he or she does not have the methods).

Having these beliefs and expectations *A* now learns that *m* is a good mathematician (**G***m*) and makes an inference from **G***m* to not **S***m*. However, if *A* learns that *m* is a good mathematician and does have the necessary methods (**G***m* and **M***m*), then this piece of new knowledge contradicts his or her present system of beliefs and expectations, and consequently something has to be given up. One can argue about what to give up. One way to solve the dilemma is to give up the belief that if *m* is a good mathematician then *m* does not find a solution (**G***m* then not-**S***m*); and the consequence that if *m* is a good mathematician then *m* does not have the methods (**G***m* then not-**M***m*). What remains is the belief that if *m* is a good mathematician and *m* has the methods, then *m* has a solution; plus its logical consequences. There is no inconsistency between this new set of beliefs and the obtained piece of information. *A* can now make a nonmonotonic inference from **G***m* and **M***m* to **S***m*.

Following Gärdenfors[6] and Makinson[7] we say that: α nonmonotonically entails β if and only if β follows logically from α together with 'as many as possible' of our expectations as are compatible with α.

Our expectations are all defeasible, however, they have different degrees of defeasibility. Gärdenfors and Makinson have proved that: if we assume that there is an ordering of our expectations, which satisfies three rather natural conditions (transitivity, dominance and conjunctiveness), then it follows that: *Nonmonotonic logic is nothing but classical logic if relevant expectations are added as explicit premises!*

Indirectly this theorem teaches us something very important. Creativity has very little to do with inferences. Monotonic inferences are as uncreative as nonmonotonic inferences. From *my perspective*, I might find your reasoning nonmonotonic and creative. From *your perspective*, however, your reasoning is in perfect harmony with classical logic — your inferences are not violating any rules of monotonicity. From my perspective, you are not following the tracks, you are making an impossible move by going West at Audley End. But from your perspective, on your map, with your expectations, there are tracks going west from Audley End. It is just that I cannot see them, they are not part of my expectations, they are not on my map of what there is.

[6] Gärdenfors (1993).
[7] Gärdenfors and Makinson (1991).

It should be obvious then that creativity has to do with the dynamics of our expectations, with changes of representation. To find a theory of (human) creativity, we have to ask ourselves what an expectation is, how it is formed and transformed. To find such a theory, let us start by looking at various theories of belief.

There are two classical theories. A mentalistic theory tells us that a belief is a mental act, or as Hume puts it: a belief 'may be most accurately defin'd, a lively idea related to or associated with a present impression' (*A Treatise of Human Nature*, Book I, Part II, Section VII). There are several well-known problems with this type of theory, but these have no relevance to the present context. A dispositional theory of belief looks at people's cognitive activities in a radically different way, in this case beliefs and behavioural consequences are linked together. But, if 'I believe that *p*', means that I entertain *p* and have a disposition to act as if *p* were true, we may get a theory which lacks explanatory value. Equating belief with behavioural consequences means that we cannot explain people's behaviour through their beliefs and desires.

There is, however, a far better alternative. The most interesting theory of belief is F.P. Ramsey's theory which says that a belief is a mental state, or, as he puts it 'a belief . . . is a map of neighbouring space by which we steer'.[8] To believe *p* is to have *p* on one's mental map, i.e. believing is a mental state. Ramsey's theory avoids the drawbacks of a mentalistic and a dispositional theory of belief while it retains their advantages. The acquisition of belief is a mental occurrence, i.e. our beliefs form a map of the world and this map can be redrawn in various ways; it can be expanded or given a greater dissolution by adding new beliefs, or it may be revised by subtractions of beliefs from it. These maps guide our actions; we steer by them in more or less the same way as we do with ordinary maps. But they cannot be equated with behavioural consequences and thus they can be used for an explanation of people's behaviour.

Attributing beliefs and desires to human beings seems rather unproblematic. Intuitively we all act as if we have a set of values and various beliefs and degree of beliefs. But, assume that our beliefs and desires are simply epiphenomena of something else, i.e. of our expectations. We do not try to phone a colleague at home instead of at work because our (aggregated) beliefs and desires tell us that this is the best alternative, but because we *expect* him to be at home. It is our value-impregnated expectations that are our basis for action. If this is correct, our mental states are far more complex than we first anticipated and what is needed is not a theory of belief and a theory of value, but a theory of expectation. Thus, paraphrasing Ramsey, one could say that an *expectation* is a map of neighbouring space by which we steer.

We have seen that if we want to understand the mechanisms behind what we call creativity, these expectations must be the object of our research. But,

[8] Ramsey (1990), page 146.

to study how people form and transform expectations is not a task for philosophers, it is something which should be done within experimental psychology. What a philosopher could do is to say what types of expectation transformation are rational; how one set of expectations should be transformed into a new set given various pieces of evidence. But since I strongly believe that not too much of interest can be said about the dynamics of expectations, I will leave the topic.

The psychologist, however, can teach us a number of useful things: How do we get our expectations? How do we transform them? How do we keep them consistent, if they are? If we receive evidence that contradict our expectations, how do we change them?

But what can an assortment of examples and arguments tell us about creativity? If a creative person is but one who has power to create, most of us are creative. But what most of us bring into being is not that novel. Nor do we want the adjective to be reduced to being used just as an honorary title or degree. Instead I have tried to emphasize what marks a more radical and far-reaching form of creativity.

Thus, a computer, whether it be one that makes drawings or a jazz improviser, is as creative as a pianola.[9] A listener might well find the music original, rule-breaking, and creative; but for the one who built the machine, it is just following a set of well-known rules. Similarly, using illustrations from computer graphics and computer music as paradigms of creativity tends to confuse the issue. If a computer produces truly novel jazz, it is not the computer or its program that is creative, it is the person who wrote the program that stood for the creativity. A computer that produces what experts take to be a previously unknown piece of music by Mozart or a drawing by Lloyd Wright is but a well-trained parrot.

What I have said can be reformulated in terms of so-called conceptual spaces. Moving around in a conceptual space is generally a highly uncreative activity. It often consists in redrawing and refining the map. It is, however, the shift from one space to another that is the sign of profound creativity. And what characterizes such a shift is the creation of new concepts, new expectations — and with it a basic change of values.[10]

Department of Philosophy
Gothenburg University, Sweden

---

[9] Boden (1994) seems to take the opposite view.

[10] A question one might want to ask is: Do we want people to be creative? How many truly creative, rule-breaking and value-transforming persons can a society or an organisation take before it collapses? Or, to rephrase the question, how many creative accountants can a bank tolerate? It seems obvious that in order to function properly a society or an organisation has to develop effective tools and methods to suppress creativity. Cf. Smith's article in this volume.

REFERENCES

Boden, M.A., *The Creative Mind: Myths and Mechanisms*, Weidenfeld and Nicolson, London 1990.

Boden, M. A., "What is creativity?" in *Dimensions of Creativity*, ed. by M. A. Boden, MIT Press, London 1994.

Courant, R. and Robbins, H., *What is Mathematics?* Oxford University Press, London (1958) 1969.

Peter Englund, *Förflutenhetens landskap*, Atlantis, Stockholm 1991.

Gärdenfors, P., "The role of expectations in reasoning", LUCS 21, Lund 1993.

Gärdenfors, P. and Makinson, D., "Nonmonotonic inference based on expectations", *Artificial Intelligence*, 65, 1994, 197–245.

Mellor, D.H., "How much of the mind is a computer", in *Matters of Metaphysics*, Cambridge University Press, Cambridge 1991, 61–81.

Moore, G.E., "A reply to my critics", *The Philosophy of G. E. Moore*, ed. by P. A. Schilpp, Northwestern University Press, Chicago 1942, 553–677.

Pearsall, R., *The Worm in the Bud: The World of Victorian Sexuality*, Pimlico, London (1969) 1993.

Ramsey, F.P., *Philosophical Papers*, ed. by D.H. Mellor, Cambridge University Press, Cambridge 1990.

JAAKKO HINTIKKA

ON CREATIVITY IN REASONING

In my native language, Finnish, the word for creating, *luoda*, means etymologically "throwing" or "throwing out" or even "throwing out of the way". This etymological meaning explains why the very same word is applied to shoveling snow. Last winter, living as I do in the Boston area, I spent more time than usual in "creative" work in the Finnish sense of the word.

Creativity is a difficult idea to get a grasp of. As was just seen, even natural languages have to resort to metaphors in expressing it. These metaphors are nevertheless in need of unpacking. That unpacking in turn easily raises questions, and it is a task that requires all the conceptual resources we can muster.

Among those resources are the insights offered by contemporary logical theory. The main purpose of this paper is to emphasize and to illustrate the uses of such logical insights for the purpose of the study of creativity at least in the area of creative reasoning. It seems to me that these insights have been neglected in much of the recent work on creativity, to the detriment of the level of argumentation and theorizing. Instead, many recent approaches use the resources of computational modelling and in some cases cognitive psychology. The use of ideas and concepts drawn from those fields is compatible with the use of results from logical theory, but in my judgment cannot replace them.

Let me illustrate these points by specific examples. Perhaps the most general problem on which some light can be thrown from the vantage point of logical and epistemological theory concerns the relation of creativity to the notion of *rule*. It is not clear what this relation is, or is intended to be, by different researchers. In (1990, p. 40), Margaret Boden writes:

> A merely novel idea is one which can be described and/or produced by the same set of generative rules as are other, familiar ideas. A genuinely original, or creative, idea is one which cannot.

Such statements are hard to reconcile with much of what creativity theorists like Boden actually say and do. For one massive fact, many of their example and applications concern the discovery of mathematical theorems, often via the study of mechanical theorem-proving in artificial intelligence. Now the formulation and the proving of theorems in an established mathematical theory of the kind creativity theorists typically are considering takes place by means of well-established rules, the same for all theorems. There is no obvi-

Å. E. Andersson and N.-E. Sahlin (eds.), The Complexity of Creativity, 67–78.

ous difference between the rules by means of which "merely novel" theorems and "genuinely original" theorems are codified and derived from the axioms of the theory in question. Both sets of rules are the same. (But see below about shortcut rules of reasoning.)

In the actual argumentation of creativity theorists, distinctions are made in terms of the mental operations associated with the different steps of the logical argument, instead of the actual deductive rules relied on in the step in question. (Cf. Boden, op. cit., pp. 108–110.) This trades on an ambiguity of the notion of rule, and will be argued below to be beside the point, anyway.

What is needed here is a distinction between two kinds of rules. The distinction is appreciated most readily in the case of games of strategy like chess. What are usually called the rules of chess I will call *definitory* rules. They specify how chessmen may be moved on the board and, more generally, what counts as a legal move in chess, what counts as winning, losing, and a tie. Such rules in an obvious sense define what chess is. If a player tries to make a move in contradistinction to the definitory rules of chess, it does not count. It has to be taken back, as if nothing had happened. All this is easily generalized to other goal-direct activities, including logical proof and (rule-governed) reasoning in general.

The other side of the same coin is that the definitory rules of a game are merely permissive. They do not say anything as to what a player should do or about which moves are good or bad or better than others. If you only know the definitory moves of chess, you cannot even say that you know how to play chess. For the purpose, some elementary knowledge of what counts as a good or a bad move. Rules to that effect will be called *strategic* rules. From game theory we know that such rules can in the last analysis be formulated only in terms of what in game theory are called strategies. Strategies in this game-theoretical sense are rules that specify what a given player should do in every possible situation that can arise in the course of a play of the game. In practice, such strategies in the strict sense are too complicated to be actually usable. Their players must rely on sundry partial strategies. The conceptual distinction nevertheless remains.

The implications of my distinction for the study of creativity are now obvious. Definitory rules are not directly connected with creativity at all. All moves are made in accordance with them, creative as well as noncreative. Creativity is a matter of strategic rules in all the human activities to which my distinction applies. But they are not move-by-move rules. In principle, they always involve longer sequences of moves.

This is the reason why we can speak, as creativity theorists do, of creativity also in the case of activities like theorem-proving which are always governed by the same rules, in creative proofs as well as in noncreative ones. These rules are definitory rules and hence irrelevant to questions of creativity. These questions pertain to the strategy a logician uses in constructing his or her proof.

These observations have several important consequences. One thing that can be seen here is that the so-called rules of inference of deductive logic are merely definitory. They are not "laws of thought" either in the sense of laws governing the way people actually reason or in the sense of how they should reason. At best, they merely help to define "the game of logic". And various incompleteness results make even this role of the so-called rules of inference dubious.

A second observation pertains to a much-used and much-abused term in this area, the term "heuristic". What is meant by a heuristic rule is clearly a *strategic* rule, albeit often merely tentative and partial. As such, there is little to object to this use of the term. However, there is a danger lurking in this usage in that some discussions in the literature seem to rely on a dichotomy between ordinary (definitory) rules of inference and heuristic rules. The intended implication of this dichotomy would be that only definitory rules can be strict, that is, that strategic rules are always "heuristic" and hence only rough-and-ready. Against such misconceptions, it is to be emphasized that strategic rules can in principle be quite as sharp as definitory ones. There is an important difference of another kind, however, that seems to be relevant to questions of creativity. Definitory rules are pointless unless they are recursive, that is, unless their applicability to each given case is decidable. In contrast, it can be shown that in several interesting cases optimal strategies (strategic rules) are not recursive. In some obvious sense, playing such a game successfully requires creativity. I will not try to spell out this sense here, however.

In the light of these preparatory comments, what can be said of creativity in deductive reasoning? This question might seem to belong to the psychology of deductive reasoning. Unfortunately, little help can be expected from those quarters. The sad truth is that the entire current psychology of logical inference uses a seriously distorted perspective on the entire subject. This distortion is based on an insufficient appreciation of current systematic logical theory. What I mean is illustrated by the great theoretical novelty in the study of logical reasoning due mainly to Johnson-Laird some twenty years ago. This novelty consisted in the "discovery" that mental models play an important role in logical inferences. There has subsequently appeared a rich crop of studies of the role of such mental models in different kinds of deductive reasoning.

I have no reason or desire to contest this "discovery". What nevertheless is in order to point out is that this industry betrays psychologists' neglect of such fundamental techniques of deductive reasoning as Beth's (1955) *tableau* method (which is merely a mirror image of Gentzen's sequent calculus) or the Hintikka-Smullyan tree method. (See Smullyan 1968.) What such techniques show is that from a suitable vantage point the entire ground floor of deductive reasoning, the so-called first-order logic, is nothing but operating with certain models or approximations of models. These models can be thought of as being mental, or they can be taken to consist of sets of formulas on paper—or in this day and age perhaps rather on the screen and in the memory of a computer. In fact, from this perspective all rules of "logical inference" obviously involve "mental models". Johnson-Laird's discovery hence does not ultimately pertain to the psychology of logic. It pertains, however confusedly, to the nature of logic itself. *The most basic deductive logic is nothing but experimental model construction.* In order to see whether G follows logically from F, we try to imagine or to construct a description of a "world" (model) in which F is true but G false. If and when such an experimental countermodel construction leads inevitably to a dead end (read: explicit contradiction) in all directions, we can be assured that the logical implication from F to G does obtain.

Hence the proper question for psychologists like Johnson-Laird to ask is not what role mental models play in deductive reasoning. The right problems to raise here include the following: What role is played by methods other than experimental modelling in the psychology of human deductive reasoning? What are these other methods of reasoning like anyway? Why is the role of mental modelling so easily overlooked, even though it is the basic mode of logical reasoning? What is the relation of the different methods to creativity?

Here systematic logical theory again provides some answers. What is crucial here, according to what has been found, are the strategies of experimental (counter)model construction. Such strategies mean in practice anticipating where the model construction leads us and above all guiding the course of the model construction. And as far as this second matter is concerned, the most important way of controlling the process of construction is to select the right new individuals to be introduced into the model. Such an introduction takes place by means of existential instantiation, for instance by means of a step from

(1) $(\exists x)S_i[x]$ to

(2) $S_i[\alpha]$

where α is a new individual constant ("dummy name"). It can be thought of as standing for an imaginary individual introduced into one's attempted model construction in the step from (1) to (2).

In general, at any stage of an attempted deduction, construed as an experimental model construction, there are several existential statements (1) present to which a step of existential instantiation has not yet been applied. The crucial question in the choice of one's deductive strategy is to determine in what order they are to be treated, including which such existential formula is to be instantiated first. For instance, the question as to how soon one's attempted deduction succeeds depends essentially on these choices.

This is all that needs to be said of the introduction of new individuals in a tree method. In a "double-book-keeping" method like Beth's *tableau* method (Beth 1955) universal instantiation in the "false" (right-hand) column operates just like existential instantiation in the "true" column—or in a tree method.

In the most general treatment of the logic of quantification, the introduction of individuals via existential instantiation must be replaced by an introduction of functions which express the dependence of the choice of the new individual on others. Such functions also serve as a vehicle of introducing new individuals via being applied to different arguments.

This important role of steps of existential instantiation in (tree-method type) logical reasoning has a well-known manifestation in geometrical reasoning. In axiomatic elementary geometry, most of the reasoning takes place by means of first-order logic. There the steps of existential instantiation are traditionally known as auxiliary constructions. Their role is crucial in the success of a proof of a typical theorem in elementary geometry. (See here Hintikka 1974.) This role did not escape the attention of many traditional philosophers of mathematics. For instance, Leibniz complained that in geometry "we still do not have a method of making the best constructions". And Kant saw in the use of auxiliary constructions as well as in the preliminary instantiation (or *ekthesis*) the source of the synthetic character of mathematical reasoning.

Obviously, one important kind of creativity in deductive reasoning lies precisely in the right choice of the existential formulas to be instantiated. This creative character is reflected in many different phenomena. For instance, there is no mechanical (recursive) method for determining the optimal choice for all situations. If one could predict merely how many such existential instantiations are needed in a proof, one could predict the minimum length of the desired proof; and so on.

The way of applying the other (tree-method) rules can affect the length and the simplicity of the proof, but it does not crucially affect the course of the thought-experiment that an attempted proof is. Such applications of the other rules basically mean only that one subsumes the configuration of individuals already reached under the general conditions on the desired model. Applications of these other rules can be in principle be made mechanically. Clever choices of how these rules are applied can significantly simplify the resulting proof and make it more "elegant". This cleverness nevertheless is of a kind that can be programmed into a computer.

Proofs in which existential instantiations are not used (or are used only with respect to existentially quantified formulas present already in the premises) constitute thus an especially simple special case of logical arguments. The difference between such arguments and those which depend on the introduction of genuinely new individuals caught the eye of as early a thinker as Charles Sanders Peirce who once called the distinction "my first real discovery in logic". (See Hintikka 1983.) He borrowed traditional geometrical terms for the distinction, calling reasoning with merely old individuals "corollarial" and reasoning that involves the introduction of new ones "theorematic".

These observations can be illustrated by applying them to one of the favorite examples kicked around in the AI literature. It is the proof of Proposition 1, 5 of Euclid which was not only produced by one of the early theorem-proving programs (see Gelentner 1963) but which turned out to be much more "elegant" than Euclid's. (For details, see Boden 1990, pp. 105–111.) I was probably the first one to point out to Marvin Minsky (private communication) that this neat proof had been anticipated by Pappus by some 1650 years. Undeterred by this precedent, this proof has been repeatedly paraded in the AI literature as a proof of the deductive skills of computers.

In the light of what has been said, the Euclid-Pappus example is a singularly unconvincing example of the alleged creativity of computers. The Pappian proof is, in Peirce's terminology, a merely corollarial one. In it, the use of Euclid's auxiliary constructions is avoided by a simple application of symmetry considerations. These considerations are of the kind that can be captured by recursive rules. They amount to manipulating conceptually a configuration of geometrical objects previously introduced as distinguished from amplifying that configuration by adjoining to it suitable new individuals.

There is no reason to deny that such use of symmetry principles as is instantiated by Pappus' argument involves some degree of creativity. However, this degree is a very low one. We tend to confuse, when applauding Pappus' ingenuity, with each other one particular application of symmetry principles (which is what is at issue here) and the idea of deciding to being such principles to bear on the problem in the first place. The latter decision is what strikes us as being clever here. But a second look at those symmetry principles shows that the strategic ideas they involve are relatively superficial ones.

What would be mistaken here is to praise Pappus at the expense of Minsky's theorem-proving program. It has been suggested that the creativity of Pappus' line of thought is due to the fact that it involves actual mental manipulation of the given figure. Pappus as it were cuts out the given triangle, turns it around, and fits it back into its old slot inverted. Isn't this more creative than a computer's preprogrammed following of its rules of operation?

It is important to realize that this contrast between a human being's actual mental manipulation of a given figure and a machine's automatic operation in accordance with its rules is a spurious one. What I have pointed out is that all logical argumentation in a suitable first-order system of deduction (roughly speaking, in a natural deduction system) not only can but must be interpreted as a gradual construction of certain mental (or on-screen) models and a series of manipulations of them. This interpretation does not depend on any heuristic imagery that we humans might resort to as a helpmeet of our psychological processes. It is a fact about the logical argument itself. Objectively speaking, a third-century Greek geometer and an twentieth-century

American computer guru must be said to be both indulging in an experimental manipulation of certain configurations of geometrical objects. Any difference in this respect is a matter of degree and detail not of the kind that would support an exclusive claim to creativity on the part of the human. I can imagine a child cutting out a Pappian triangle and accidentally fitting it back into the same slot like an inverted jigsaw-puzzle piece, without thereby exhibiting any ingenuity in geometrical theorem-proving. I can also imagine you watching a screen on which a computer is printing line by line the Pappian proof in a suitable notation and at a certain point exclaiming: "That's ingenious! The computer has inverted the substitution-values in that universal-quantifier formula". Any black-and-white difference in creativity between a human being and a computer in this particular case is based on an optical illusion. What is crucial is the use of a certain symmetry principle, not the imagery one associates with it.

The mistake I have criticized cannot even be excused by pointing out that very nearly the entire philosophical community in effect committed it at one time. When the likes of Frege and Russell developed for the first time an explicit formal logic which was powerful enough to capture the modes of reasoning used in elementary geometry (or at least most of them), they went overboard and rejected all conceptualizations that had been made in terms of the figures that traditionally accompanied geometrical proofs. These figures were seen merely as ways of introducing illicit appeals to geometrical intuition through the back door. This is among other things undoubtedly the reason why Peirce's important distinction between theorematic and corollarial reasoning was not understood and was in fact almost totally overlooked.

In reality, geometrical figures are best thought of as a fragmentary notation for geometrical proofs alternative to, but not necessarily intrinsically inferior to, the "purely logical" notation of formalized first-order logic. Being alternative notations for the same modes of reasoning, they inevitably share many of their crucial features. Among other things, as I have explained, suitable forms of purely formal logical reasoning can be thought of as involving mental experimentation with configurations of individuals quite as much as the forms of reasoning that employ figures or other "visual aids".

Such experimental model-building and mental manipulation of configurations of objects is not a mere heuristic device, a technique to facilitate deductive reasoning psychologically. They are intrinsic features of certain deductive methods. They are a part of the semantics of logical reasoning, not only of its psychology or its heuristics. If it is suggested that heuristic ways of thinking are needed to make mathematical reasoning intuitive, I will borrow a line from Wittgenstein's *Tractatus* 6.233 and say that in this case the language (notation) itself provides the intuitions. Generally speaking, my results put the entire discussion of mental representation in cognitive science to a new light. If language itself provides the requisite representations, what is the problem?

Constructing model-like thought-experiments and manipulating them is what deductive reasoning is. It is not *per se* a mark of creativity. It is part and parcel of the problem, not of its creative (or noncreative) solution.

Thus Russell and his ilk were wrong. The formalization of a line of reasoning that originally utilized figures or other model-oriented devices does not necessarily mean that conceptualizations that were formulated by reference to the figures lose their meaning. Among other things, if the formalization is of the right sort, we can still speak of the introduction of new individuals (e.g. new geometrical objects) into the reasoning and of other kinds of manipulations of certain configurations of individuals. In particular, it can be said in general that the most creative aspect of deductive reasoning in general is the choice of the new individuals to be introduced into the reasoning. Choosing the things one tries to establish about individuals already introduced is an important problem in practice. Without good guidelines such an enterprise soon leads into complications beyond the actual control of the most powerful computer. But the problem of managing this enterprise does not lead us beyond what can be done by means of fixed mechanical rules.

Why, then, has the nature of attempted logical inference as experimental model-building been so thoroughly misunderstood? The answer is again forthcoming from systematic logical theory. What natural-deduction rules (in the sense of the expression used here) give us is the simplest, most basic and (surprise, surprise) most natural method of deduction. All the other methods actually used can be thought of as arising from natural deduction methods by allowing certain shortcuts. From proof theory we can even know what the most general form of such shortcuts can be taken to be. It consists in adding to the list of formulas that are supposed to be true in the model under construction a tautological disjunction of the form (S∨~S), where the choice of S is up to the ingenuity of the logician. This rule is sometimes known as the cut rule.

From the famous result known as Gentzen's first *Hauptsatz* (see Szabo 1969) it follows that whatever can be proved by means of the cut rule can also be proved without it, by mens of the original natural deduction rules. What the cut rule can do is to shorten the proof. But choosing the optimal applications (yielding the shortest proof possible) of the cut rule cannot be specified by any recursive (mechanical) rules. Indeed, there is no way of finding recursively the shortest proof of a given logically true formula even if it is known to be provable. I will call ingenuity in choosing the best applications of the cut rule creativity of the second kind.

How does this agree with the original idea of an attempted deductive proof as an experimental (counter)model construction? What an addition of (S∨~S) to the list of formulas to be made true in the eventual model (if any) means is that the construction is split into two branches, one geared to satisfying S and the other to satisfying ~S. In each branch, there is a new formula present which may be of any complexity whatsoever. Such a formula can shorten to construction (proof) by facilitating the introduction of new individuals through existential instantiation.

In effect, instead of introducing new individuals one by one into the argument, the deductive reasoner introduces in one fell swoop a store of many such individuals. This wholesale introduction of new individuals is what facilitates the simplification of the proof. The possibility of this simplification depends on the choice of S and hence of the reasoner's ingenuity.

But what is introduced in the first place in an application of the cut rule is not a number of individuals directly. What is introduced is a sentence which allows the repeated introduction of such new individuals. The sentence itself is merely a conduit for the introduction of individuals. The sentence is discursive, not iconic. It is not a part of the model under construction. When it is introduced, the reasoner has to consider what it says about the world at large, not how it contributes to the step-by-step construction of a putative counter-model.

Thus it is the shortcut rules like the cut rule that in a sense violate the idea of logical deductions as experimental model constructions. They are the element in deductive reasoning that is discursive, in contradistinction to the picture-like model constructions that natural-deduction proofs are.

Here we can see how the conventional idea of logical reasoning as a discursive process has come about and also see the reason why it involves a fundamentally wrong idea of deductive reasoning. The natural deduction rules (in effect model construction rules) are so natural that most of the time they are applied automatically, without the reasoner's even being aware of doing so. What he or she is aware of, sometimes painfully aware of, is the need of clever applications of shortcut rules. And these shortcut rules are discursive and not merely steps in a mental (or nonmental) model construction or manipulation of partial models.

From the vantage point of the natural deduction logic which can so directly be interpreted in terms of model building, the shortcut rules like the cut rule achieve their shortcut effect by introducing a number of new objects into the model construction wholesale. But these new objects are not introduced iconically, as new building-blocks added to old ones. Instead, they are introduced by linguistic (symbolic) descriptions of certain kinds of states of affairs. What objects they introduce has to be figured out from these symbolic statements. This is what explains their discursive character as contrasted to the direct mental modelling effected by natural deduction methods (or at least implicit in them).

This diagnosis fits well with the selection by psychologists and philosophers of what they consider paradigmatic logical rules. The most frequent candidate for this role is undoubtedly *modus ponens*. What it enables one to do is to prove G by proving F and $F \supset G$. It can be considered as a weakened form of the cut rule which authorizes us to prove G by proving that neither $\sim F$ nor $(F \,\&\, \sim G)$ has any models. *Modus ponens* is therefore a mere shortcut rule, not part and parcel of the basic rules of deductive logic. Its discursive character makes it a misleading example of how the basic rules of logic really operate.

In general, the apparently discursive character of deductive logic is thus an optical illusion. At bottom, all deduction is experimental model construction, but to facilitate such model constructions (and to keep them under control) we resort to discursive methods which have stolen the show from the more basic natural deduction rules.

These results enable us to identify several different kinds of creativity in deductive reasoning. It might seem that the use of shortcut rules is to be considered as the most important form of creativity. The shortest possible natural deduction proof of a logical truth T is like the par for a golf course. It is not the limit of what a logician has to do in order to prove T effectively. It only marks a score which a good conservative logician can achieve. The real skill (creativity) is according to this view shown by a logician's using shortcut rules to simplify the proof, just as a superior golfer can break the par of a course. Creativity in deduction is exhibited by an efficient use of the shortcut rules.

This idea is not implausible. The application of the shortcut rules leaves much more scope for the ingenuity of the logician than the natural deduction rules. In fact, there is no mechanical (recursive) method of finding the shortest proof when shortcut rules are allowed even when it is known that the theorem in question is in fact provable. (In contrast, such a method does exist for natural deduction proofs.) The shortcut rules therefore leave plenty of room for insight, intuition, and serendipity.

As far as deductive reasoning is concerned, however, the shortcut rules must in my judgment be considered only as a way of anticipating the kind of introduction of individuals (and functions) on which the real success or failure of an attempted deduction rests. The basic form of creativity in deductive reasoning lies in being able to anticipate in some way or other the situations to which a logician's experimental countermodel construction leads. And such an anticipation is basically a matter of anticipating what new individuals there are to be introduced when.

The kind of ingenuity exhibited by the Pappus-Minsky proof is not of this kind. It deals only with experimentation with already constructed configurations and is therefore less important (less creative) than the anticipation of the outcome of the introduction of new objects.

Now the interrogative model of reasoning and argumentation which I have developed allows me to extend what has been said so far from deductive reasoning to all reasoning and argumentation almost verbatim. (See here Hintikka et al., forthcoming.) The gist of all creative reasoning lies in anticipating the outcome of the introduction of new individuals into the argument and *a fortiori* in the introduction of the right ones. If an example is needed, Sherlock Holmes' "curious incident of the dog in night-time" will serve the purpose. What is Mr. Holmes doing here? Everybody in the story "Silver Blaze" where this example comes from has been puzzled about the nocturnal disappearance of the famous racing horse and the apparent murder of its trainer, the stable-master, out in the heath in the same night. The reason why they are puzzled is that they are considering only the principals of the story, the horse, the stable-master, the thief, and the killer. (Of course these need not be all different from each other.) Holmes introduces a new ingredient into the configuration, viz. the watch-dog, and almost at once part of the mystery is solved. Did the dog bark when the horse was stolen? No, it did not even wake up the stable-boys in the loft. Now who is it that a trained watch-dog does not bark at in the middle of the night? Its owner, the stable-master, of course. Hence it was the stable-master who stole the horse ... Holmes' feat of reasoning is just like that of a geometrician who proves a theorem by means of an appropriate "auxiliary construction", that is to say, by introducing a new object into the argument.

Likewise, Peirce's distinction between theorematic and corollarial reasoning can be extended to all (interrogative) reasoning.

The main technical difference is that now new individuals and functions can be introduced into reasoning by an oracle's answers to the inquirer's questions. This *mutatis* is nevertheless an obvious and natural *mutandis*.

Also, the role of shortcut rules is now taken over by arbitrary yes-or-no questions. Even though in general we must require that the presupposition of a question has been established before it may be raised, the presuppositions of yes-or-no questions are trivial and can therefore be overlooked. This actually increases the range of conclusions that can be established, unlike the use of shortcut rules in deduction.

In general, the same things can by and large be said of creativity in deductive reasoning and in reasoning in general, as it is captured by the interrogative model. This is connected with the fact that there is a very close connection between the choice of optimal strategies in deductive reasoning and the choice of such interrogative reasoning as aims at the discovery of new truths, as shown in Hintikka (1989). (Cf. also what was said above about the connection between creativity and strategic rules.)

The conclusions reached here have implications for all discussions of creativity in reasoning and more generally of creativity in rule governed activities. I will leave these implications for another occasion—or perhaps for the reader to work out himself or herself.

Department of Philosophy
Boston University, USA

REFERENCES

Beth, Evert, 1955: "Semantic Entailment and Formal Provability", *Mededelingen van de Koninklijke Nederlandse Akademie van Wetenschappen*, Afdeling Letterkunde, N.R., vol. 18, no. 13, pp. 309–342.

Boden, Margaret, 1990: *The Creative Mind: Myths and Mechanisms*, George Weidenfeld and Nicholson, London.

Gelentner, H., 1963: "Realization of a Geometry-theorem Proving Machine", in Edward A. Feigenbaum and Julian Feldman, editors, *Computers and Thought*, McGraw-Hill, New York, pp. 134–152.

Hintikka, Jaakko, 1974: *Logic, Language-Games and Information*, Clarendon Press, Oxford.

Hintikka, Jaakko, 1983: "C.S. Peirce's 'First Real Discovery' and Its Contemporary Significance", in *The Relevance of Charles Peirce*, ed. by Eugene Freeman, The Hegeler Institute, La Salle, pp. 107–118.

Hintikka, Jaakko, 1989: "The Role of Logic in Argumentation", *The Monist* vol. 72, no. 1, pp. 3–24.

Hintikka, Jaakko, forthcoming: *The Principles of Mathematics Revisited*, Cambridge University Press.

Hintikka, Jaakko, Halonen, Ilpo, and Mutanen, Arto: "Interrogative Logic as a General Theory of Reasoning", forthcoming.

Hintikka, Jaakko, and Koura, Antti, forthcoming: "An Effective Interpolation Theorem for First-Order Logic".

Johnson-Laird, P.N., 1983: *Mental Models*, Cambridge University Press.

Smullyan, Raymond, 1968: *First-Order Logic*, Springer-Verlag, Berlin-Heidelberg, New York.

Szabo, M.E., editor, 1969: *The Collected Papers of Gerhard Gentzen*, North-Holland, Amsterdam.

DONALD G. SAARI AND ANNELI L. SAARI

TOWARD A MATHEMATICAL MODELING OF CREATIVITY

"Creativity" is fascinating! We know so much about the topic without having the slightest idea what it is. We even know how to promote creative behavior by encouraging subjects to avoid strict adherence to rules, to explore new explanations or paradigms, to welcome novelty, to transform a problem into a different framework, to generalize through abstraction, to brainstorm, and to explore options before making evaluations. But, other than knowing that these approaches tend to work (sufficiently well so that some are "corporate brainstorming strategies" while others are tried in the classroom), we really don't know *why*. The "why" becomes an important issue. Also, all these different approaches emphasize "generality" over details. What explains this commonality?

Continuing, researchers have uncovered clues suggesting that creativity is associated with a complex dynamic. This is manifested by those "thinking is not thought" phrases. The purpose of these catchy comments is to focus on the immense gap separating where conscious and dedicated acts are used to resolve a cognitive conflict from some sort of sub- or preconscious rearrangement of ideas. This subconscious notion includes the important and not uncommon "aha!" experience where a difficult problem suddenly is resolved during a period of rest, upon awaking, or while taking a walk rather than during a period of active work. Well known examples of these almost religious happenings are described in Mozart's famous letter, in Poincaré's often quoted lecture, and in Hadamard's book (1945). This also is part of Helmhotz's "inspiration" stage that Poincaré includes in his outline of creative thinking. (The Helmhotz and Poincaré stages have been rediscovered by many others.) As this preconscious activity is an important creativity characteristic, we need to understand why and how it happens.

"Triggering", where seemingly unrelated events unleash an idea, is another fascinating creativity phenomenon. A red car driving past may trigger a solution to a financial problem; overhearing a casual conversation may trigger, in mysterious ways, the resolution for a difficult scientific problem. Why?

The mystery continues. When pressed for an explanation of creativity, standard responses invoke "intuition". But, what is intuition? Typical answers

The research for this written version of a talk presented at the *Cognition and Creativity Workshop* in Venice, Italy, October, 1994, was supported by an NSF grant and the Arthur and Gladys Pancoe Chair in Mathematics.

Å. E. Andersson and N.-E. Sahlin (eds.), The Complexity of Creativity, 79–103.

strive to capture the sense that "intuition is knowing without knowing why". Clearly there is something called intuition, but we have no real notion of what it is or how it works. To understand creativity, we need to understand intuition.

Abstraction

So, we have a large store of knowledge *about* creativity and the creative process without really understanding what it is. A way to resolve this paradox is to establish a general framework to unite what we understand about the creative and cognitive processes while providing a structure from which to find new relationships. Our approach differs from the literature in that we abstract standard cognitive traits into a *dynamical model* and then, rather than allowing issues from cognition and creativity to dictate our assumptions, we let the mathematics determine which questions to explore and what added structures to impose.

Dynamics is not new for the exploration of creativity; e.g., see Arieti (1976), the equilibration notions of Piaget, Szekely (1976), Erickson (1977), Kragh and Smith (1970), Smith and Carlsson (1990), and many others. Other "creativity" approaches, starting over a century ago with Galton's work (1874), emphasize personality traits including tenacity and the ability to work hard. For still others, the definitions and emphasized concepts intentionally mirror methodological approaches, or they rely upon anecdotal information for support. For several, "correctness" of creative output is an important factor.

While we agree that personality and other traits help characterize the creative individual, we ignore them primarily because they are not consequences of the mathematics. Also, we find them to be analogous to describing the driver of a racing car; some traits represent idiosyncrasies such as the brands of sun glasses and driving gloves preferred by skilled drivers, while others represent skills needed to activate and coordinate the vehicle. But, a car's potential is governed and constrained by the *car's* features. After all, even a skilled driver performs differently in a 1960 VW and a 1997 Porsche. Similarly, hard work, tenacity, and the ability to dream in Technicolor does not guarantee creativity; they describe what may be needed to harness and activate the available cognitive options. Our emphasis is to understand the available options and mechanisms of the "creativity car".

By adopting a deeper level of abstraction than commonly found in psychodynamics, we find that certain creativity traits can be explained solely in terms of the *adaptation* of cognitive processes. Our abstraction emphasizes the structural elements supporting "creativity", independent of whether it is a human or an organization, so it may help us understand what promotes a "creative society" or a "creative organization", whether "creativity" can be encouraged in the classroom, in organizations, and in society, or whether governmental policies frustrate, rather than assist, the development of a creative society. (To illustrate, consider the continual conflict between the need for expression and newly imposed standards in our schools. When forced to prepare for new tests and curricula, teachers occasionally are criticized that their students are not as creative.)

This essay describes portions of a more general model about cognitive processes. The selected traits (from an extension of (Saari, 1978)) provide plausible explanations for the "aha" experience, intuition, similarities in "creativity enhancing approaches", and "triggering". Because our assumptions are governed by the mathematics, not cognition, we are encouraged that several conclusions resemble common behavior. But, to keep our exposition from becoming overly abstract, intuitive descriptions replace a careful, formal presentation. Appropriate terms, however, are defined in the footnotes so that the reader with a mathematical background can supply the missing details of our plausibility arguments.

TOWARD MODELING

Our model is simple and sparse; it uses minimal, accepted assumptions about thought processes. Essentially, we just assume that, somehow, information is conveyed from the outside world to the attention of the individual, the organization, or the computer. From this barest of structures—the properties of transmitting information—we extract relationships and consequences. By using standard assumptions, our results extend to (but may not have been noticed) other models.

As stated, our assumptions are *not* motivated by the cognition or creativity literature; they are dictated by the *mathematics*. At each stage, the *mathematics* requires selecting from several options; we explore and describe the consequences of certain paths. So, those abstract conclusions which resemble creative and/or cognitive properties identify which assumptions explain such activities for people, organizations, or computers. Much of our discussion describes parallels between the abstract conclusions and results from the cognitive, creativity, and learning disability literatures.[1]

We emphasize *adaptation*—a central, widely accepted concept in psychology used to model "adjustment". Our objective is to find a "cognitive adaptation" foundation to unite and explain a wide variety of seemingly disparate creative and cognitive behavior. In this way, we determine which of the tacit assumptions of this field already admit creativity-like behavior. As the modelling involves the adjustment of information and as similar adjustments occur in society, organizations, and even computer programs, there exist a wide spectrum of applications and ways to test our conclusions.

[1] We view comparisons with the cognitive and the LD literatures as a partial test of our assumptions.

Adaptation has to occur somewhere, so the mathematics requires specifying the underlying space. Treat this space as a sterile "black box" that could represent a person, an organization, a computer, or whatever is being analyzed. Differences in derived consequences depend upon the properties of each black box.

World views

There are many novel ways the "world" can be viewed; our model must admit all of them. This is important; if creative people view the world differently than us more mundane beings, then their insights must be represented. In fact, this issue needs to be confronted if we hope to address the concern that "many people assume that there will never be a scientific theory of creativity—for how could science possibly explain fundamental novelties?" (Boden, 1994, p. 75) Fortunately, the mathematics demands a wide variety of differences. As these "world-views" form the foundation for the more interesting "creativity" discussion, we devote considerable attention to this issue.

Piaget does not state the "world view" issue in the above manner, but he does emphasize the central nature of the problem by labelling the way a person interprets the world as her "Organization"; this "Organization" is important for Piaget's theory. This term with a similar meaning appears throughout the literature. For instance, with Kragh and Smith's (1970, p. 27) approach which places a heavy emphasis on methodology, "[t]he concept of *organization* will denote the visual form reported by the subject verbally and/or in the form of a drawing".

Following standard usage "environment" refers to the outside world. The environment's actual organization (e.g., behavior determined by physical laws, gravity, or imposed by "our" interpretation), however, need not agree with a person's Organization. "Correctness of interpretation" may be central for creative product approaches and needed in a lab to distinguish between clever and nonsensical insights of an experimental subject, but it is counter to an "abstract" study of creativity because it imposes assumptions before they are needed. (Insisting on accuracy requires imposing an all-knowing being to determine "correct answers". But, if creativity includes an ability to develop new paradigms, then creative interpretations differ from accepted views. So, what is "correct?" As "correctness" changes, one must be suspicious of models requiring agreement between a person's and "our" Organizations. Instead, conflict (as defined by the *individual*) should drive the creative process. Again, this is consistent with the literature; Erickson, for instance, relates a person's development with his interaction with society.)

Understanding occurs somewhere, so treat the center of "thought processes" as a nondescript "*black box*", denoted by *BB*, where information from the world is conducted to *BB* through input nodes. This just resembles the standard computer metaphor for information processing. Our refinement comes from examining the consequences of encoding information.

So far *BB* only possesses input nodes to convey information from the outside world. If these nodes represent, for instance, our senses, then one node might transmit visual information, another might convey sound, still another might register taste. Should *BB* represents a computer, the input nodes have a standard "hard wire" interpretation. When *BB* models an organization, each participant in the information and/or decision processes serves as a node. We need to determine how different ways to transmit information affect understanding.

Interpretation

Borrowing from Piaget, "understanding" an event is characterized by how an individual organizes or arranges the environment. As indicated, Piaget calls this structuring "Organization", In our setting, the input nodes convey information or stimuli from the environment E. The definition of E is determined by (and can change with) the choice of *BB* and the input nodes. For instance, while E need not include a Saturday football game if *BB* is a computer, it can when *BB* represents an individual. To repeat, *the structure of E is essentially ignored.* (How the E structure affects "feedback" is an important part of the more general model that is not described here.)

Instead of allowing the consequence of what is transmitted to rattle around *BB*, assume each input, or event e, has a unique output in *BB*. By treating the jth node as a mapping

$$f_j : E \to BB, \qquad (1)$$

a natural definition for *BB*'s "understanding" of event $e \in E$ is the image $f_j(e) \in BB\, f_j(\mathrm{e}) \in BB$.

This functional relationship captures minimal aspects of most, if not all, theories of cognitive development. After all, it just asserts that, somehow, information is brought to *BB*'s attention. For instance, Eq. 1 corresponds to the input node of the computer metaphor, it represents basic aspects of Piaget's "assimilation", and it captures minimal parts of Smith's "Construction in the direction of the object" stage in his Precept-Genesis (PG) analysis (see (Smith and Carlsson, Fig. 2.1, p. 17)). For other illustrations, one node for a mathematician may correspond to an analytic representation of a problem, while another may be a geometric formulation. More loosely, think of an "interpretation" (the image of the mappings) as the retrieval and association of a current event with past experiences.

Already mathematics poses questions to explore. For instance, most mappings identify many inputs with the same image.[2] For Eq. 1, this requires several events to share the same interpretation. This, of course, captures the

[2] Generically, functions are not one-to-one.

human behavior where the same meaning can be associated with each of

"peculiar", "pecular", "*peculiar*", "PECULIAR", or "*peculair*"

even though the second and last choices are misspelled and different typesetting is used. So, an immediate abstract consequence of Eq. 1 corresponds to a well recognized cognitive trait. (In fact, a curiosity is how conclusions from most mathematical assumptions admit illustrating examples.) In highly specialized situations, mathematics allows input nodes to have the property where each $b \in BB$ represents a unique input $e \in E$[3]. If this property models a computer programed to recognize only special fonts and correct spelling, then only one "peculiar" choice may be accepted. As we show later, abstract arguments suggest that this precision has a cost; intuition is lost.

For another implication of Eq. 1, if two input nodes differ, even very slightly, then they have different interpretations for at least one input. Again, this property has close parallels with human and organizational behavior; e.g., we know that a variety of events can be equated with hearing something while a different collection might be equated with seeing it. Carrying this comparison of different functions a step further, it allows individuals—even with similar wiring—to have different world views. After all, difference in the nodes requires events which lead to different interpretations. Thus, just by assuming Eq. 1, the model permits even seemingly similar individuals or organizations to exhibit different Organizations; this meets our goal of admitting a wide spectrum of Organizations.

Organization

To extract further Eq. 1 consequences, observe that *BB*'s "Organization" are all events sharing the same interpretation; this is for all choices of b in *BB*. Thus *BB*'s Organization critically depends upon the choice and properties of how information is transmitted. To mathematically define "Organization", recall that the inverse function $f_j^{-1}(b)$ identifies all possible inputs—all possible events — with b as the output (the interpretation). As $f_j^{-1}(e)$ is the *BB* interpretation of e, $f_j^{-1}(f(e))$ is the set of all events sharing the same interpretation as "e". This is illustrated in Fig. 1 where the interpretation of event, e, is depicted by the arrow pointing to the right, $f(e) = b$. The two dashed arrows identify other events from the environment, $f^{-1}(f(e)) \in E$, with the same interpretation; i.e., f cannot distinguish among them. (Compare this statement with the figure and the "peculiar" example.) "Organization", then, is the structuring of the environment as determined by these inverse sets.

---

[3] The mappings are one-to-one.

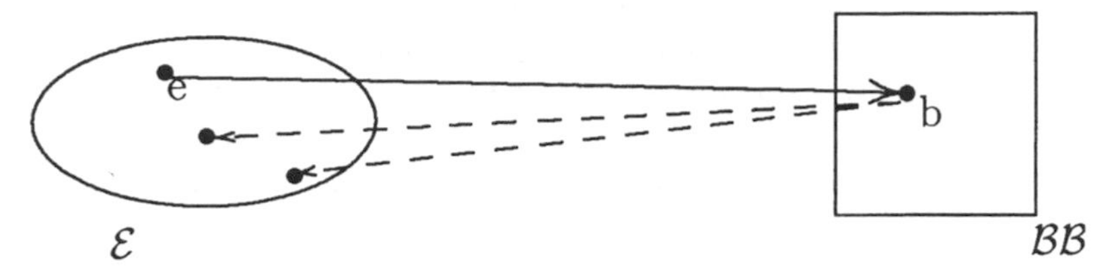

Figure 1. *Organization; events with $f(e) = b$ interpretation*

This Eq. 1 modeling requires an "interpretation" (the image element b from BB) to be based on the properties of the entries of BB. If, for instance, BB models a computer that just lists physical objects, identification is the only allowed interpretation. However, should the BB modeling admit relationships, then b could be an admissible relationship involving, perhaps, addition or multiplication. If, for instance, e is the event of tossing three apples in with four apples then $f_j(e) = b$ might be the abstract "3+4=7" addition process and $f_j^{-1}(f(e))$ might represent other "3+4" addition problems involving cars, or strawberries, or . . .[4] Then again, the transmission of information (the input node) might only associate fruit with addition. Namely, it follows from the mathematics that determining whether events can be interpreted, how they are interpreted, and how they are associated with other events critically depends upon properties of the input nodes and the BB structure. Notice the variety of world views! So, one way we introduce a variety of ways to view E is by including all ways to transmit information. Compare this inclusive approach with the tendency—almost an obsession—of other models to specify exact properties. This specification, or "fixation", imposes unnecessary obstacles to understanding creativity.

Creativity

This "different people view the world differently" structure sheds light on a mystery of the creative process. We have in mind Hadamard's (1945, p. 49) worry about "the failure of a research scholar to perceive an important immediate consequence of his own conclusions". How could this be? How could a bright person who is the world expert on a new concept by discovering it, miss an important, "obvious" consequence? As Hadamard shows with examples, this is not uncommon.

A plausible explanation uses the observation that different functions have different inverse sets. So, treat a new discovery as an input e. For me, e is associated with all events $f_j^{-1}(f(e))$; for you and your input node g, the same

[4] Nothing in this modeling requires awareness of all possibilities. For instance, we don't know all events that can be combined, but if they involve addition, we still would use arithmetic.

event is associated with all events $g_j^{-1}(g(e))$. Differences in how we view the world, as represented by different input nodes, require us to have different associations with the same e. So, even if I understand the idea I invented better than anyone else, our differences of associations, modeled by each of us having different input nodes, allows you to make associations with e that I failed to see; you could find implications I missed. A similar description holds for organizations. "IBM started the personal computer revolution in the early 1980's and then was almost undone by it because it did not grasp the impact of what it had set in motion. When customers began to abandon their expensive IBM mainframes for clusters of cheaper personal computers, Big Blue was unprepared. Its revenues plunged, . . ."[5]

What does this abstract structure suggest about developing creativity? The most obvious lesson is to avoid adopting identical world views. In order to generate different associations—and potentially different insights—the structure of *BB* and/or the input nodes must differ. In terms of a person or organization, this suggests avoiding becoming a carbon copy of someone else. Indeed, successfully mimicking someone is modeled by assuming both have essentially the same Eq. 1 functions. As the inverse sets—the associations—closely agree, it becomes difficult for an imitator to discover something new through different associations.

In practical terms, this supports J.E. Littlewood's advice "[d]on't think you must read up all the literature that *it might* has a bearing [on a new and difficult mathematical problem]". To develop a creative approach, Littlewood (1968) advises that "there is much to be said for going ahead . . . without reading anything beyond . . . the minimum to find what the problem is about". Littlewood's suggestion is consistent with what is suggested by our abstract structure. Remember, the input nodes represent how information is transmitted. So, a researcher carefully reading the literature—beyond understanding the problem—runs the risk of representing, formulating, and transmitting information about the problem in a way similar to the giants in the field. Even children know better than to compete with giants on their terrain with their weapons.

Simple examples

The assertion that radically different Organizations emerge with different choices of *BB* and/or how information is transmitted (the input nodes) is so critical that it is worth illustrating how the conclusion comes from the model rather than an imposed wish. To do so, we invent easily analyzed "black boxes" which have nothing to do with humans or organizations.

Our test site has E as the set of all real numbers—positive, negative, and zero—and *BB* as the set of non-negative numbers. One mapping to convey

[5] New York Times, 12/18/95.

information is $g(x) = |x|$; it interprets the world of numbers, E, in terms of a number's magnitude as g ignores the sign. This "input node", therefore, cannot distinguish between -1 and 1; i.e., $g^{-1}(g(1)) = g^{-1}(1) = \{1, -1\}$. Thus the world according to g (i.e., g's Organization) combines numbers into pairs according to magnitude. (Compare this description with Fig. 1.) Contrast g's world view with that of the mapping h which drops the integer portion of a number while retaining the fractional part. As h's Organization is defined strictly by fractional parts, this input node prohibits *BB* from distinguishing between 4.5 and 6.5. Indeed, as $h(4.5) = .5$, it follows that $h^{-1}(h(4.5)) = h^{-1}(0.5)$ is the set of all numbers with fractional part 0.5. So, g and h have the same interpretation of 0.5, but h detects no difference between 0.5 and 100.5 while g finds a major difference. Notice how differences in transmitting (formulating, etc.) information can significantly alter interpretations.

To show that even slight differences in input nodes can generate major differences in perception, treat h as the remainder of a number when divided by 1. A closely related choice is the function k which specifies the remainder of a number when divided by 1.1. While h and k agree on the interpretation of the input 0.99, they disagree about 1.01; h views 1.01 as the insignificant 0.01 while k interprets it as the large value 1.01. Thus, even slight changes in input nodes can create significant differences in interpretations!

This mathematical example underscores the sensitive dependency of the Organization, or world view, upon the properties of the input nodes. With the same *BB*, radical differences in the world views of g, h, and k result from the different ways they convey information. This general consequence of Eq. 1 occurs in all settings.

As for creativity, this structure suggests that imaginative outcomes need not always be the result of personality, drive, or smarts. It suggests that even a seemingly dull person can, at times, appear to be imaginative. Again, even closely related functions admit different Organizations. So, while different individuals or organizations usually have essentially equivalent interpretations, any slight differences could allow situations with very different interpretations! While such differences may correspond to bankruptcy or another disaster, they also might allow a person identified with mundane views to have a creative interpretation in certain settings. Is this creativity? It is if the outcome is a new view. So, this statement, which depends on what associations are made, suggests that some types of "creativity" may be due to circumstances. (The movie *Forest Gump* cleverly creates settings where an undistinguished person excels.)

Similarity

Once an Eq. 1 relationship is specified, the next mathematical step is to worry about the "topology" of the image space. This requires determining whether and how *BB* accepts two interpretations as being almost the same. To illustrate, our *BB* consisting of non-negative numbers defines two numbers as being close or similar when their difference is "small", The "peculiar" example includes the slightly misspelled words.

By adding this similarity structure to *BB*, the "Organization of *E*" inherits a sense of "closeness"; inputs are similar should they admit "nearby" *BB* interpretations.[6] In other words, the mathematics provides a working definition for "associated events"—a key concept from cognition and creativity.

Illustrating with our mathematical examples, the input node g treats -5 and -5.01 as being essentially the same because g's respective interpretations of 5 and 5.01 differ by only a hundredth of a unit. While this seems obvious, remember that "closeness" is determined by how information is transmitted. To illustrate, while *we* don't accept -50 and 50.0001 as being near each other — they differ by more than a hundred units — g does. By interpreting these events, respectively, as 50 and 50.0001, g finds them to be closer together than -5 and -5.01. This example underscores the (Eq. 1) fact that interpretation is in the eye of the beholder.

Notice how this example illustrates our earlier comment that *BB*'s Organization of the universe need not be accurate, nor reflect what someone else accepts. Instead, it is influenced by the input node. For instance, while g judges 2.3 to be closer to 2.1 than to 9.11, h disagrees. Because h interprets 2.1, 9.11, 2.3 as being, respectively, 0.1, 0.11, 0.3, h treats 9.11 as being closer to 2.1 than to 2.3.

The assumption that the *BB* structure defines a sense of "closeness" or proximity provides an even richer variety of "Organization" structures. Differences in the assumed *BB* structures (e.g., different educations, cultural backgrounds, or upbring), introduce even more "world views". This makes sense; it captures the sense that different life experiences can cause different interpretations of events. (To illustrate with the test site of *BB* and function h, let "closeness" in *BB* now be defined by the remainders when a number is divided by 10. This forces 9.99 and 10.01 to be far apart because the first has a remainder of 9.99 while the second has a remainder of 0.01. Therefore, the new *BB* structure combined with k now renders 9,99 and 10.01 as being distance, while the former structure had them close to each other.)

Thus, changes in the structure of *BB* and/or how information is transmitted can alter *BB*'s world view. This preliminary portion of the modelling already quantifies and explains those "People think differently because they are different" type of statements. In particular, it now is easy to extend the earlier "creativity" comments concerning Hadamard's concern about why an inventor of an idea need not recognize important applications. Different life experiences define different definitions of "proximity". In turn, this can create radically different associations and Organizations. Consequently, we

[6] Thus "*E* has the weak topology induced by $\{g_j\}$".

must expect the same stimulating event to have very different meanings and associations with different people.

Aggregated interpretations

Different *BB* input nodes create conflicts in interpretations. This is because each mapping defines a preferred Organization of *E*, and, by being different, these Organizations cannot always agree. This conflict needs to be resolved.

Nonabstract examples of this conflict are easy to find. It already occurs with our sensory input nodes where one is hearing and the other is sight. "Seeing" a small portion of a red vehicle, for instance, might generate a wide equivalency class (that is, a large number of possible interpretations and associations) ranging from a flashy sports car to a fire truck. Hearing a siren brings forth interpretations ranging from a police to an ambulance to a fire truck. The final interpretation resulting from aggregating information gathered by these senses is obvious; it is a fire truck simply because that event is common for both mappings.

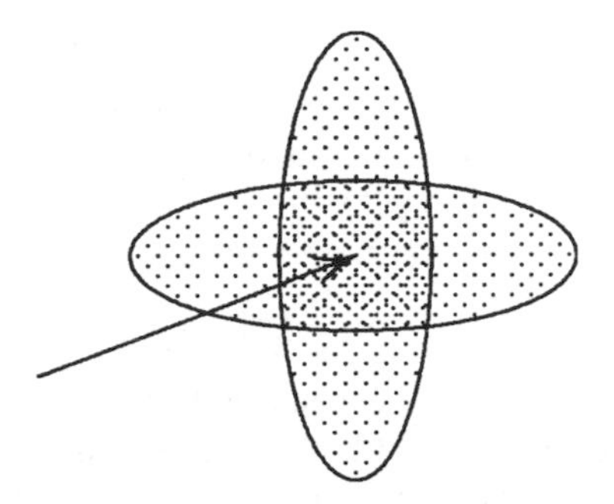

Figure 2. *Reaching an interpretation*

Similarly, each Fig. 2 oval depicts the Organizations generated by an input with each way to transmit information. Define the *BB aggregated interpretation* to be the part common to all of the input mechanisms; it is the intersection (the heavier shaded region) of the different different ovals. As an intersection includes only what is common among the Organizations of the different nodes, it represents what is understood without conflict.

To show how this definition resembles behavior in organizations, computers, and humans, start with the fact that an intersection is an exclusionary act. Thus, the aggregated interpretation can be thought of as the end result of some sort of evaluation process. While the dynamic is unspecified (Sect. 4), it has the flavor of evaluating through comparing the associations of all the input nodes. Thus the aggregated interpretation — the intersection — represents a completed evaluation.

To better appreciate this notion, suppose *BB*, normally serviced by several mappings, finds some nodes inoperative. Figure 2 dictates what must happen. As some sets (ovals) are missing, the intersection is larger. This carries the sense of a fuzzier interpretation—one that is not as refined or sharp as it would have been.[7] This fuzziness represents the unmade comparisons from the missing nodes.

With reflection, this description is as it should be; because *BB* is missing refinements afforded by other perspectives, the final intersection is not as carefully "evaluated", so it need not be as refined. This abstract conclusion admits immediate parallels with the problems of humans who are blind, deaf, or suffer a learning disability. In each setting, the deficiencies of a particular way to convey information dictates a gap in the Organization; comprehension is impaired. A related behavior is when an organization loses a member with a unique and valued perspective on organizational issues. For individuals or organizations, assistance is provided by finding how to overcome the roadblocks.

Notice the distinctly different meanings for the "size" of these regions. A large "size" of the events directly associated with a particular interpretation represents "fuzziness" of thought. This makes sense; it means that discriminations have not been made. On the other hand, a large size of events accepted *as being similar to an event* indicate a wide range of associations. These *associated events* can be equated with an individual who has an awareness of other data, a flexibility of thought process, an ability to fuller utilize other events from the environment[8]—it can be associated with creativity.

So, different distinctions of the "size" of these regions could represent "creativity" (when wide and unusual associations are emphasized), or only a partially developed sense of the world (when many undistinguished events are viewed as the same) such as a child using"truck" to identify all vehicles. While these two interpretations of a "large region" represent very different structural sources and different levels of sophistication, both require a "wide association" with a specified input. Thus we must expect "creativity" to share child-like behavior. (This connection is not original, but its explanation is.) In turn, these comments provide partial support for experts who worry whether creative-like actions in children indicates creativity, or an undeveloped sense of discrimination.

[7] The induced topology is cruder than the original one.

[8] All conflict in "size" differences disappears when these ideas are formally expressed in terms of topology. "Sharper" ideas correspond to a refined topology; abundance of associations is identified with the size of open sets.

Further differences

Significantly different "Organizations" can be generated by imposing different structures[9] on how *BB* interprets "closeness". At one extreme, *BB* may not have such a definition because every event is interpreted as being distinct from all others.[10] This *BB* can only cluster events that are equally interpreted; there is no ability to understand that events are related or close to one another. This can be illustrated with the early word processors on computers—until a notion of "proximity" for words was programmed into the computer, spell checkers were impossible because all words had to be interpreted as distinct entities.

Examples mimicking this abstract behavior are easy to find from, say, the learning disability literature. For instance, compare these "discrete topology" consequences with the observed "Executive function difficulties" in retardation. Here, the person can perceive various stimuli, but he cannot pay attention to relevant aspects of a problem—he cannot recognize the close relationship of other events.[11]

The other extreme is where "everything is close to everything else".[12] Here *BB* has different interpretations for different inputs, but "everything" is related without distinction whether something is "more similar" than something else. Immediate mimicking examples come from certain computer structures; unless programmed, the computer may be able to recognize that one string differs from another, but it clumps them all together—everything is essentially as different as anything else.

The rich variety of structures between these extremes provide a more comfortable middle ground; they allow for varying degrees of similarities. Examples resembling such behavior include the spell-checker illustration. Some spell-checkers suggest alternatives for misspelled words only if they differ by a letter; others, where the programming provides a more sophisticated notion of "closeness", offer several alternatives. Interesting related examples come from the child development literature. One example is the tendency of a child to first learn the letter "o", and then the child equates all letters with "o". You can almost see the crude sense of "closeness;" if it is a letter, it is called "o". Once differences—more discrimination—are noted, letters are called by the correct name. Notice how when the proximity regions become more refined, there is a concomitant change in the interpretations. Similarly, with the earlier "truck" example, much too quickly the child learns to accurately identify a passing vehicle as "a '57 Chevy with its classic wings, but with four-on-the-floor and eight cylinders".

---

[9] That is, by imposing different topologies on *BB*.

[10] *BB* has the *discrete topology*.

[11] So, a person without this problem is modeled with a reasonable topology; a person with it is modeled with the discrete topology where "closeness" does not exist.

[12] Here *BB* has the *trivial topology*.

TOWARD INTUITION

How do creative people discover original solutions? As usual answers involve "intuition", we need to represent this sense that we can know something without knowing it. This description transforms intuition into something that is suspected but not recognized as being relevant. Our model already admits this notion.

To explain, assume several "input-output" mappings service *BB*. (Again, when nodes are defined by our senses, there is at least one input node for each sense.) Figure 2 is a schematic representation for the expected conflict in the Organizational regions; a conflict that becomes more pronounced when the "similar interpretations" admitted by the proximity structure of *BB* are used. In Fig. 3, we describe these regions in terms of intuition.

Each Fig. 3 oval represents an input node's Organization. The heavier shaded intersection is the firm, aggregated interpretation of e. The remaining regions—depicted by the portions of the ovals outside of the "interpretation region"—represent associated events that, through at least one input node, *BB* treats as being related to e. What prevents these other events from being identified with e is that they are eliminated from the intersection through the evaluation processes—they are removed because of their conflict with associations admitted by the other ways to transmit information. So, rather than serving as an interpretation of e, events in these regions are only "suspected" of being related to e. These oval regions outside of the intersection, then, model a form of "intuition"; call them the *intuition regions*.

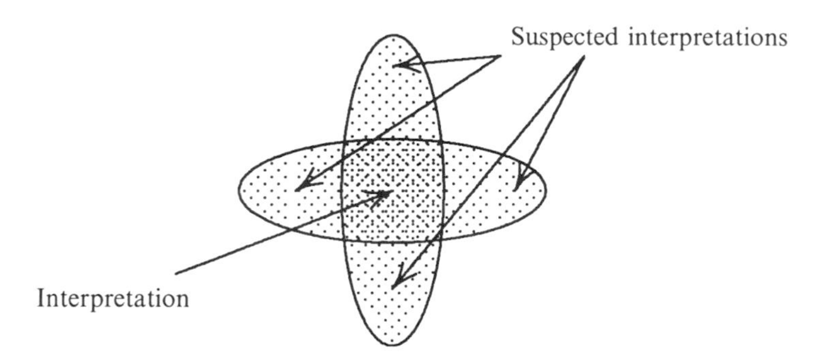

Figure 3. *Intuition*

This definition replaces the mysticism of intuition with a property that can be examined, tested, and even exploited. For instance, instead of identifying the act of discovery with mystery, luck, pure chance, or accidents, this definition anchors intuition and discovery upon previous preparation (to develop the proximity structures of *BB* and how information is transmitted). It suggests how to introduce and evaluate personality traits and means of "thinking" to facilitate creativity. (Certain traits and approaches better allow these intuition regions to be explored than others.) Not only does this comforting comment support the virtue of hard work, but it provides relief to Hadamard's concern [1945, p. 19] about descriptions concerning new discoveries. He worried that "explanation by *pure* chance is equivalent to no explanation at all and to asserting that there are effects without cause". Our structure, critically based on what *BB* knows, avoids this creativity trap.

Immediate "intuition" consequences follow from the now standard observation that different ways of transmitting information define different associations. By considering different kinds of nodes and/or different kinds of "proximity structures" on *BB*, different kinds of "intuition" emerge. So, different people, different organizations, etc., must be expected to have different types of intuition.

To start our analysis, consider how information is transmitted. An extreme setting has a single input that requires precision. As a single input does not admit an intuition region, "intuition" must reflect the "similarity" structure of associated events. However, if this similarity structure is missing (the precision assumption), then so is intuition. An illustrating example is where a computer requires each interpretation to hold only for a single event with zero tolerance for errors (i.e., there is no definition of proximity). Although this precision permits rapid computations, it also empties the intuition regions of Fig. 3. Thus precision without similarity kills intuition! "Intuition" can be partially re-introduced to computers through the programming of "heuristics"; these are "rules of thumb" identifying what events are similar to other events. Thus, heuristics can be viewed as attempts to recapture aspects of the Fig. 3 intuition.

For another example, consider Piaget's general and discriminatory assimilations. Assuming a "similarity" structure and postulating the existence of nodes with these properties requires one Fig. 3 oval to represent all events that are similar to an input in terms of general features, while the second oval emphasizes details. The intersection, or region of general agreement, is where both are satisfied. As the remaining regions represent types of intuition, one intuition region emphasizes events which agree in general terms while the other concerns events which agree in certain details (say, in details of a proof).

Extending this analysis to organizations is easy. Here the intuition regions may correspond to similarities in how a product is produced, or how it is marketed, or how it is financed, or . . . depending on the different divisions of labor and decisions.

If a person is intuitive in one area, will that person exhibit intuition in another? According to this structure, the answer is "Not necessarily". Our definition of intuition heavily depends on the "association" regions, so it is based on the particular input. The way a function is defined on a certain region of a space need not dictate how it is defined elsewhere, so the intuition regions can differ with different inputs.

Assisting intuition

This working definition for intuition suggests how to enhance "intuition" and, hence, "creative thinking". The idea is obvious; adopt procedures which encourage the exploration of each intuition region. While the exact manner this is done depends on the structure of *BB* and the input nodes, the general idea is to "turn off" the evaluation. Remember, the intuition regions are where not all critical comparisons have been made. So, momentarily turning off certain input nodes while considering implications of others allows "fuzzier" comparisons; new associations are included even though they may be contradicted by other information (i.e., comparisons with other nodes). Sound familiar?

How this is done depends on the *BB* structure. To see this, consider an organization where separate units report to a central decision unit. This division of responsibility allows charging the different divisions (the different ovals) to separately consider options without worrying whether they are compatible with options from other divisions. What makes it easy to describe intuition searches for organizations is that evaluation stage can be turned off and on. The same is true for any *BB*, even individuals, allowing such a decentralization. This "turning off" approach is captured by the common refrain heard in a mathematical research session of "For the moment, forget about the fact that . . ."

Intuition searches become more difficult with *BB* choices where the evaluation is not separate. In particular, for humans, rather than being conscious acts, many decisions are made sub- or unconsciously. (In "Adaptation", this subconscious activity is modeled by adaptation.) When this happens, a conscious act is needed to partially suspend judgement until after the various options, or "intuition regions" are explored. Notice how this description corresponds to the approaches mentioned in the introductory paragraph of this essay. Indeed,

> *brainstorming*, which requires individuals to suggest resolutions without making judgements as to their feasibility or effectiveness, clearly is a conscious manner of ignoring evaluation until after the intuition regions are explored.

Similarly, the purpose of

> *abstraction*, with its emphasis on discarding all traits that are not essential, is to enlarge the similarity Organizational regions associated with each *e*. Again, this creates larger intuition regions.

A closely related approach widely used in mathematical research is

> *simplification*, where the actual problem is replaced with a much simpler version. By removing the complexity of the original problem, more connections can be made (so, the "proximity" regions are increased in size)

and judgement about how to resolve the original issue is delayed (information about the original problem is ignored).

Other methods admit similar descriptions. For instance,

different framings, by placing an object in a different framework, change the associated emphasis of the Organization; it requires new environmental events to be in the oval regions.

Also notice how this approach supports ideas from education (e.g., trying to create the appropriate atmosphere in the classroom where risk-taking becomes socially acceptable) where the emphasis is on exploring options.

ADAPTATION

We now include adaptation. We need this notion to reflect the realism that an input $e \in E$ cannot be instantly assigned an interpretation $b \in BB$; instead b should result from a search. Of the many possible models, including trees, we examine "adaptation". Namely, we describe searches that start by "adjusting" the current situation. This assumption leads to several natural mathematical structures; a partial selection and their consequences follows.

Think of adaptation in terms of a metal ball attracted by magnets.[13] After firmly attaching several magnets, suppose a metal ball is placed on the smooth table. This ball may be attracted toward one of the magnets, but which one, and how long it takes depends on the attracting strength of each magnet and the initial position of the ball. A more interesting setting is if we had a variety of magnets where each type attracts only certain kinds of metals, but not others. This more general setting captures the sense of our modeling.

Mimicking the metal ball, adaptation requires an initial perception of an event to be refined and attracted toward the best fit of an interpretation. How long this adjustment takes depends upon the capabilities (the wiring) of BB, the input nodes, and the interpretations (magnets). Complicating the issue is that each input node adjusts the interpretation. Thus, simultaneously the interpretation of each node may be pulled in different directions (i.e., toward different final interpretations). This structure allows all sorts of scenarios. For instance, one could imagine a setting where one node is near a final interpretation when the effect of another node suddenly and radically changes the interpretation.[14] Indeed, one can even construct scenarios with alternating interpretations!

---

[13] Adaptation is modelled as a dynamical system determined by the structure of its attractors.

[14] This describes a standard hyperbolic structure where the motion moves near the stable manifold until, near the equilibrium, the unstable manifold begins to dominate.

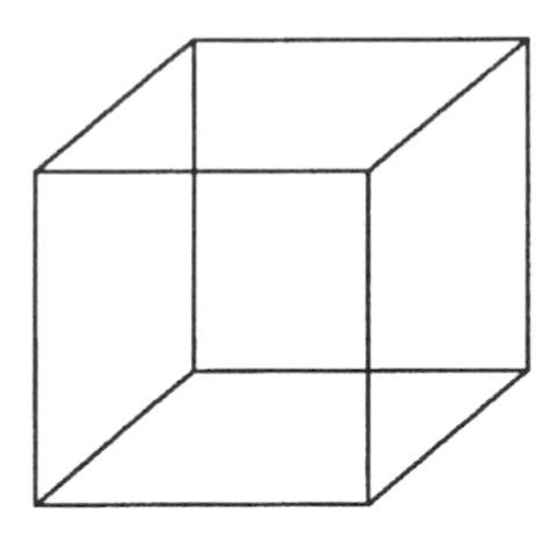

Figure 4. *Necker cube*

Admittedly, this scenario seems preposterous. To show that it is not while making these comments a tad more concrete, stare for awhile at the Necker cube of Fig. 4. The initial well-defined interpretation allows you to specify whether a particular corner (say, the vertex about the "N") is in the back or front. But, after staring long enough, the orientation suddenly reverses; the corner differs from what it was! And, the orientation changes again, and again, and again. Because sight uses several nodes to transmit information, this "spontaneous restructuring" phenomenon illustrates our cyclic scenario caused by an adaptation conflict in "final interpretations". (Each node provides a "local interpretation" of an image; the full picture requires coordinating the pieces. As the top and the bottom skewed rectangles could be interpreted as hidden or observed, the flipping occurs. By exploiting this "local interpretation" for each node, other illusions are found by creating a picture separating conflicting "local" information; Fig. 5 is another famous one.)

A more standard scenario is where a final interpretation is quickly reached. Here, the sense of "adjustment from the current situation" introduces efficiency for routine searches. ("Routine" is where what happens next is closely related (for *BB*) to what is happening now.) For such inputs, rather than engaging in a lengthy tree search, adaptation saves time by adjusting a current interpretation. Just as the initial position of the metal ball affects the final choice of an attracting magnet, this adaptation modeling introduces a near-neighbor prejudice for how *BB* interprets events. (To appreciate the *BB* dependency on near-by interpretations, let *BB* model a person's first visit to a radically different culture. Rather than quickly handling and interpreting routine events, the novelty can cause disruption.)

An immediate "creativity" implication comes from this "near neighbor" adaptation property where the interpretation of an input can vary depending on the situation. If *BB* is interpreting the environment when a new input *e* occurs, then *e*'s initial interpretation is influenced by the current status and focus of Organization. With a *BB* structure rich enough for *e* to have multiple interpretations, the final belief can critically depend on the current setting.

(This corresponds to how the current location of the metal ball determines which magnet is the final attractor.) This adjustment property, then, suggests that a final interpretation can be influenced even by seemingly superfluous background "noise". On the positive side, this captures the "triggering" phenomenon from creativity. On the negative side, a desired interpretation could be disrupted when pulled away by "noise" or distractions; e.g., recall the frustration of losing an idea by politely listening to someone else first. This dynamical argument captures the flavor of Mozart's (Hadamard, 1945, p. 16) description of a creative moment. "Then my soul is on fire with inspiration, if however nothing occurs to distract my attention". Also, this "local prejudice" suggests explanations for the "fixation" obstacle of creativity.

Figure 5. *Illusion; two or three legs?*

Speed

How quickly can an interpretation be attained? While it might be almost instantaneous with computers, the adjustment speeds of different units (i.e., different input nodes) of an organization can vary depending upon the task and the information and decision structures. With individuals, differences occur with different people and different inputs; e.g., the speed of interpretation for sight and sound differ. Thus the modeling of speed requires specific assumptions about BB, the choice of e, and the way information is conveyed to BB (the input nodes).

We take a different approach by appealing to dynamical systems for guidance about the general dynamics. Here, it is safe to assume that rate of convergence to an interpretation increases the closer it is to a non-conflicting "fit". Conversely, if an initial perception is "far" from a final, non-conflicting interpretation, the weaker attraction and slower rates of change make the final interpretation more susceptible to interpretations from other nodes and other influences.[15] (With the magnet analogy, these assumptions state

[15] Included in this instability statement is where slight changes in position force an initial interpretation into a different basin of attraction.

that the closer the ball is to an attracting magnet, the faster it moves. Conversely, a ball that is far away moves slowly. With several competing magnets, such as when the ball is in the center of a ring of magnets, a slight change in its position can radically change the identity of the attracting magnet. Identifying "magnet" with final interpretation and the ball with an "event" provides a sense of the assumptions while suggesting conclusions.)

Reaching interpretations

Adaptation occurs in *BB*, so each adjusted *BB* position represents a momentary interpretation of *e*. Remember, each node pulls this short-lived interpretation toward its preferred attractor. Thus, similar to a flip of the Necker cube, a momentary interpretation acceptable for one input node but not another, can be changed.

We now can resolve the mystery of how a final interpretation is reached. An interpretation is dragged out of an intuitive region by the adaptive force of a node that finds it conflicting. This continual refinement of the transitory interpretations finally settles at an interpretation reasonably acceptable for each means of transmitting information. (The final resting position could be a compromise point; e.g., a metal ball placed halfway between two magnets remains there. Elsewhere we show how these positions cause new interpretations.) Namely, the adaptation dynamics creates the aggregated interpretation of Fig. 2! On the other hand, adaptation drives momentary interpretations from the "intuition regions", so it is adaptation that must be partially countered for "creative actions". (This, we believe, is where "personality traits" may play a major role.)

As an application, recall the common creativity trait, related to the "aha" experience, where first the general outline, but not the details, of an idea are discovered. This is described in Mozart's letter where "[o]nce I have my theme, another melody comes, linking itself to the first one, in accordance with the needs of the composition as a whole . . . The works grows; I keep expanding it, conceiving it more and more clearly . . . It does not come to me successively, with its various parts worked out in detail . . ." A similar emphasis on the general theme is when Poincaré, describing an important mathematical discovery, states "I did not verify the idea; I should not have had time . . . On my return to Caen, . . . I verified the result at my leisure". With so many other examples, it is reasonable to accept that the general theme precedes the details. Why?

One explanation emphasizes the competition of input nodes; each way of transmitting information excludes certain interpretations during the adjustment process. Thus, different adaptation speeds, or the tendency to rest at one interpretation before moving to another (as manifested by Necker's cube), generate dynamical behavior resembling these descriptions. Alternatively, if one applies our adaptation assumptions to Piaget's notions of generalizing and discriminatory assimilation, where the former moves more rapidly than the latter, then the dynamics requires the general picture and theme to be followed by the details.

Behavior during adaptation

Before exploring other adaptation consequences, we need to interpret the changing meanings in terms of the rate of adjustment. With a rapid adaptation, these fleeting interpretations are quickly replaced by the final interpretation. Here, the momentary interpretations are not much of a factor; an individual or organization may not even notice them. A slower adjustment, however, requires lingering at each transient interpretation. As these interpretations represent disequilibrium and conflict, they may be manifested by a sense of confusion where "confusion" depends on the characteristics of BB.

How long this "temporary confusion" lasts depends on how alien e is from the current focus. (With magnets, this corresponds to their strength and the position of the ball.) If the new e is close to the current BB focus (where "close" is in terms of the BB structure), then the adaptation assumption prejudicing the initial search in favor of a current focus requires a quick interpretation. Again, this captures the ease of interpreting daily events. Conversely, the more alien e is from the current focus of BB, and the slower the adaptation process, the longer the confusion time. An example illustrating this slower effect where BB is an organization is when you made an unusual request for an airline ticket, or a nonstandard order at a restaurant. Events within normal operating procedures are easily dispatched. But, depending on the efficiency (i.e., the adaptation speed of the organization decision and informational structures), an event out of the ordinary can create difficulties.

For individuals, inputs fitting within a common structure are quickly assimilated; those that differ from the current focus take longer to comprehend. A common example is when you quickly and unexpectedly encounter an acquaintance in an unusual surroundings—say, your neighbor from Chicago walking out of a Stockholm restaurant. Here the sense of confusion can even be manifested by momentary dizziness. Another example comes from the belief that certain LD difficulties reflect a slower adaptation rate. Support for this hypothesis comes from an effective teaching technique that is recommended for classrooms with both "regular" and "LD" students. Rather than the usual approach of calling on a student and asking a question, the order is reversed and a time gap is introduced. Namely, after a question is asked, and a short time span (allowing for adaptation), a student is called upon. Our adaptation modeling supports the observed positive reaction.

Environmental change

If environmental inputs can change, we must expect the relative difference in environmental and adaptation speeds to play a critical role in interpretations and creativity. If environmental events change more rapidly than the adaptation process, then the interpretations have to blend together; different events are interpreted as being connected or essentially the same whether or not they are. For instance, a movie projector projects a series of still pictures faster than our adaptation speed, so the still pictures become connected into what we perceive as a continuous motion. But if the projection speed is slowed, the continuous illusion is replaced by individual frames; events intended to be connected are not. The faster adaptation rate, then, forces this natural clumping to disappear—events need not be interpreted as being related. At the other extreme, events moving too fast for the adaptation rate, such as a movie in "fast forward", cause a blur and confusion—as too many events are being connected, general interpretations are unavailable. Examples illustrating both kinds of behavior come from the LD literature.

These comments suggest that portions of creativity result from having the correct levels of adaptation. If adjustments are too slow, everything is a blur and relationships are missed. If they are too fast, relationships are separated.

Impressions

To test this modeling, suppose an event *e* is "observed"[16] throughout the adaptation. Here, we would expect the final interpretation to be as accurate as capable with *BB*. This makes sense; *e* remains available for comparisons. Call this final outcome the *stationary interpretation* of *e*. Now suppose *e* is removed before the adaptation is complete. As *e* is not available for comparisons, the resulting interpretation is based upon the initial, partial interpretation. Call this interpretation the *impression* of *e*.

To compare how the impression and stationary interpretation can differ depending upon the circumstances, suppose an input *e* is highly alien to the current *BB* focus. Our representation of adaptation requires the initial search for an interpretation to be based on the current focus. So, if the input is quickly withdrawn, the mathematics requires a wide gap between its "impression" and its "stationary interpretation". Moreover, by not keeping *e* in the region of current attention, the initially slower search should be manifested by confusion.

There are many supporting examples resembling this theoretical prediction; e.g., it is not uncommon after an accident for a person to be unable to recall details. Or, as described earlier, recall that short period of confusion

[16] By remaining in the domain during the search for a dynamic, it provides a second dynamical force which is equivalent to changing the initial conditions.

when you unexpectedly encounter an acquaintance in a different country. A more interesting example is the widely discussed "eye-witness" phenomenon. Typically, a test group is quickly shown a picture of, say, a subway scene depicting a well dressed black business man being threatened with a knife by a ill-dressed, unsavory white man. With surprising regularity, when the picture is shown only momentarily, the race of the victim and assailant are interchanged—something that does not occur should the picture be shown for a longer period. This behavior, of course, is consistent with the description of "impression". Personal prejudices, perhaps personal experiences, determine the "nearest magnet" for this input. When it is not there for consultation and comparison, the wrong interpretation is reached.

Related examples come from trying to identify a person from pictures. The dynamics suggest that the observed pictures should distort the memory of the actual person; forensic artists are beginning to understand this fact. This, too, is consistent with prejudice of adaptation to the current Organizational focus.

Thinking is not thought

It is this connection between adaptation and its dynamical prejudice toward the status quo that provides insight into some of those "thinking is not thought" types of behavior. Because "thinking" involves abstraction, treat the space of conscious thought as the environment E where the inputs are individual thoughts. Thus, for this discussion, "thinking" is a conscious change of the E inputs. The rate of change of these inputs corresponds to "environmental change". Accompanying these change is the adaptation process which tries to eliminate conflict by reaching a final interpretation.

Our comparison of "thinking and thought" involves different, potentially conflicting dynamics. Conscious thinking, is the act of changing or holding fixed a particular e (that is, changing or fixing an idea). The second, a sub- or preconscious activity, is the adaptation procedure which tries to reach a nonconflicting interpretation. Predictions about resulting behavior comes from our analysis of the interaction between environmental and adaptation rates of change. The adaptation prejudice favoring the status quo allows events (thoughts) which are minor modifications of current interpretations to enjoy a smoothly working procedure. Here, the manipulated inputs ("thinking") lead to reasonably quick, smooth interpretations. But if an input (the process of thinking) is far from an accepted interpretation, the adaptation process (thought) is slow. Here the environmental changes dominate the adjustment (adaptation). Therefore, rather than reaching an interpretation, "thinking" actually hinders the adjustment process from reaching a final interpretation. This is true whether the consciously manipulated inputs are rapidly changing, or held fixed. But if environmental changes (thinking) are placed on hold—because the person is taking a walk, sleeping, attending a play or concert—the conflict between environmental change and adaptation is removed. This allows adaptation to search for an interpretation. By the properties of the dynamic, the closer adaptation comes to an interpretation, the faster the convergence. The dynamic, then, should be manifested by a person suddenly and unexpectedly arriving at a conclusion! In other words, this adaptation dynamic captures, in a natural fashion, the "aha!!" experience.

(Using the magnet description, identify the manipulation of conscious thought with the conscious movement of a loosely held metal ball. If the moving ball approaches a magnet, it will be pulled away by an interpretation—a magnet. However, if the ball is moved far from any magnet, it is doomed to remain in a confusion zone. This particular dynamic corresponds to fixing on a particular topic—the loosely held ball is kept in a region far from any magnet. Now, suppose even in this region of weak attraction, the hand is removed. *If* there is an attracting magnet, even if the attraction is weak, the ball will move in particular direction. Without interference, the moving ball eventually reaches a stronger attraction zone where it suddenly moves rapidly to a magnet. Aha!)

This "aha" prediction follows directly from the interaction between the two types of dynamics. This is not mystical; all of us have experienced it. All of us have suddenly, when not trying, finally remembered the name of a person, or movie, or address that earlier we could not recall. This same trait is accorded a higher status when it produces a "creative product". For instance, compare this theoretical construct with Poincaré's comments that "the change of travel made me forget my mathematical work . . . At the moment when I put my foot on the step the (important mathematical) idea came to me, without anything in my former thoughts seeming to have paved the way for it". (In Poincaré's speech reported in reference (Poincaré, 1952).) This is the "aha!!" experience.

So, behavior surprisingly reminiscent of the "aha" behavior is a natural prediction of the conflicting dynamics of consciously manipulated thoughts and adaptation. Moreover, this explanation removes mysticism. Replacing the popular "light bulb" where a powerful idea suddenly occurs out of nowhere, the dynamics requires the basis to exist. For instance, this description does not permit either of us coauthors to wake with a grand, new idea in chemistry. Instead, it requires the basis for a new idea to already be in *BB* the ideas cannot occur without preparation.

An advantage of this model is that it suggests ways to enhance this "aha" creativity experience. To introduce an "adaptation period", take a walk, go swimming, get involved in some activity where the mind is not consciously trying to manipulate the outcome of an idea, and let adaptation do its job. Because adaptation is highly dependent upon an initial position, this dynamic also suggests doing a little work or thought about an aspect of a problem just prior to an adaptation period; that is, just prior to going to a play, concert, to sleep, on a walk or drive.

This strong dependency on the choice of e suggests still other strategies. Returning to the story of the metal ball, it sure would help if the ball is moved into a region where attraction is more stronger and more likely. From our earlier discussion, this suggests exploring the different intuition regions to increase the likelihood of success of the adaptation dynamic. A natural strategy is to reformulate a problem. By definition, a reformulation introduces other associations in Fig. 3. Or, rather than trying to find new associations with a problem, find associations with an *associated problem.* For instance, by simplifying a problem, the actual problem is related to a much simpler model. This simpler model may allow new and/or extended intuition regions.

SUMMARY

The many seemingly different and sophisticated models for cognition, creative activities, and organizational behavior hare certain traits. Somehow and in some manner, information from the outside has to be conveyed and it cannot be instantaneous. By analyzing these assumptions, we discover behavior that is surprisingly similar to many of the mysteries of cognitive processes and creativity. By using standard assumptions, these consequences must be subtle, hidden properties of other models!

This particular essay provides a flavor of this type of analysis. Rather than a complete story, only particular aspects of a more general model are described with suggestive examples and description.

School of Education and Social Policy
Northwestern University, USA

Department of Mathematics
Northwestern University, USA

REFERENCES

Arieti, S., *Creativity: The Magic synthesis*, Basic Books (New York), 1976.

Boden, M., "What is Creativity?", in *Dimensions of Creativity*, (Boden, M.), MIT Press, 1994, pp. 75–118.

Erickson, E., *Toys and Reason*, Norton (New York), 1977.

Galton, F., *English men of science, their nature and nurture*, Macmillan (London), 1874.

Hadamard, J., *The Psychology of Invention in the Mathematical Field*, Princeton University Press, 1945.

Kragh, U. and Smith G., *Percept-Genetic Analysis*, Gleerup (Lund, Sweden), 1970.

Smith, G.J.W. and I.M. Carlsson, *The Creative Process*, International Universities Press, Inc. (Madison, Connecticut), 1990.

Littlewood, J., *Some Problems in Real and Complex Analysis*, Heath (Lexington, Mass), 1968.

Poincaré, H., "Mathematical creation", *The creative process*, (ed. Chiselm, B.), University of Calif. Press (Berkeley), 1952.

Saari, D.G., "Cognitive development and the dynamics of adaptation", Northwestern University preprint, 1978.

Szekely, L., *Denkverlauf, Einsamkeit und Angst*, Huber (Bern, Switzerland), 1976.

ANDERS KARLQVIST

CREATIVITY. SOME HISTORICAL FOOTNOTES FROM SCIENCE AND ART

The clue to discovery is not in the substance of one's idea,
but in what is learned from the style of one's attack.

— N. Mailer: *The armies of the night*

INTRODUCTION

Creativity is the capacity to create new things. It is a concept which embraces more than just imagination, ingenuity or orginality. It actually deals with *making new* things happen. Don Quijote and Robinson Crusoe were both imaginative men but we would certainly choose Crusoe as the creative person. Creative people are not dreamers. They do things. The philosopher Whitehead has described creativity as the formative element "whereby the actual world has its character of temporal passage to novelty". Creativity is the drive that pushes reality forward.

It is difficult to find a definition which is scientifically precise and at the same time captures the ordinary use of the word creativity. It is more easy to recognise creativity when we see it than to define this elusive quality. So a pragmatic approach to the issue would be to look around for creative people and try to discern common traits or general patterns. Such observations must however be interpreted with great care. There are no straightforward conclusions. (Rather it resembles interviews with very old people trying to find out why they managed to reach such a high age. Some would answer it is because they have kept away from smoking and drinking and spent time on physical exercise. Others would no doubt respond that they have always taken a glass of wine and smoked a good cigar after dinner.)

While still recognising the inherit difficulties in making this kind of studies and generalising from accounts from exceptional people in science and art who have proven to be creative beyond any doubts, I like to take my point of departure in some well known biographies. The first example is Isaac Newton.

Å. E. Andersson and N.-E. Sahlin (eds.), The Complexity of Creativity, 105–114.

ISAAC NEWTON

Isaac Newton was born on Christmas day 1642. His father had passed away already three months before the boy was born. The Newtons was a family of some wealth but with no education. Isaac's father could not write his own name at time when he died. His mother remarried three years after the child was born and as it turned out Isaac's stepfather would not accept the boy in their home and he was sent away to spent most of his early years with the Ayscoughs, the family of his mother. In this environment he encountered different values and it was taken for granted that he should have a basic school education. The Ayscough family also played a critical role later when Isaac was admitted to Trinity college at Cambridge. And the the story of one of the greatest achievements in the history of science could begin in ernest.

Isaac's cousin, i.e. his father's brother's son, was brought up under similar circumstances as Isaac would no doubt have faced if not his own father had died so early. His cousin died illiterate.

The step from having a creative potential to actually be able to develop this potential and make use of it, depends on lucky circumstances. In Newton's case it was obviously a narrow escape from a rural life in oblivion in a 17th century English village.

ALIENATION AND CONCENTRATION

In retrospect we can discern various factors that might have been relevant for Newton's creative development. Without digging deep into the psychological aspects and trying to evaluate the traumas that probably affected Newton in his early years—loosing his father, being separated from his mother and so on—it is evident that he was a very lonely boy, you could say an alienated person. He seems to have been at odds with his teachers and had difficulties to establish close relations with other pupils during his school years. He obviously spent a lot of time reading and he assimilated a great amount of theological as well as scientific litterature early in life. In many respects he was a true autodidact, especially in mathematics.

Newton's style of investigation is of some general interest. As just noted he read a great deal and he read with a very inquisitive mind. He asked questions. Actually he collected all these in a notebook "Quaestiones". This book contained a long list of specific queries which then became the focus of his efforts. By the process of answering these questions he gradually transcended the limts of scientific knowledge as it then existed. His scientific endevours is characterised by a tone of constant questioning.

Another quality which is apparent in Newton's research is his intensity and concentration. When asked how he had discovered the law of universal gravitation he replied: "By thinking on it continually". Another quotation along similar lines reads: "I keep the subject constantly before me, and wait till the first dawnings open slowly by little and little into a full and clear light". This quality of intense and sustained concentration has been acknowledged by many in the scientific and artistic field. Henri Poincaré gave more or less the same answer when he was asked how he made his discoveries in mathematics: "By thinking about them often". Another example (of course!) is Ludwig Wittgenstein whose struggling with the foundation of logic and with thinking itself demonstrates this virtue of concentration to its mental and physical extremes.

ISOLATION AND INTERACTION

How does this "principle of concentration" apply in a modern society? What forms can it take and what alternatives do we see? Learning from these (and other) examples it seem as if new ideas and products would need a time of incubation during which the creative process is shielded off from external influences and disturbances and when ideas are so to speak protected from each other. It is clear that classical academic environment like the colleges at Cambridge could provide such a splendid isolation. You could live and work more or less like in a monastry. (For Wittgenstein this was still not enough, he had to find his own isolated place in the Norwegian wilderness).

To confine yourself to the closed world of Trinity college or similar places does not necessarily mean that you are cut off from the rest of the world. We can see from the correspondence that Newton had that he was in close contact with many leading scientists in Europe, and it quite clear that some of these contacts had a profound influence on his work. His contacts with Edmond Halley, who visited Newton in Cambridge in August 1684 probably completely changed the course of Newton's life and thereby laid the ground for his masterpiece—the Principia. Would his work have benefitted if he had had access to e-mail? I doubt it!

Sverker Sörlin, professor of the history of ideas at Umeå university, has demonstrated convincingly in his book *De lärdas sällskap* (The Learned Society) has the international scientific network is not a new invention. Carl von Linné, the famous Swedish 18th century biologist, is a good example of such a network builder. However, it is important to note that these contacts were seldom truly interactive in the sense we talk about interaction in the computer era of today. Newton was informed , critisised and challenged but he took his time to digest and think. Often his responses were delayed several months and often his correspondence was done through a third person.

In spite of the primitive technical means communication in the 17th and 18th century it should be remembered that the intellectual elite of Europe nevertheless had a common language of communication—latin, which later came out of fashion and only recently have been replaced by English as a lingua franca for scientific communication.

The world today is different. We have means for instanteneous interaction across the globe. Information is spread immediately. Access is in practice unlimited. And the expectation to be able to access new information without delay is already part of our value system. It is more or less a technical imperative that we should be able to know immediately and to be able to react at once. Isolation and temporal delays are viewed as mere anomalies.

This is not a selfevident blessing. The almost perfect and instantaneous spread of information might prevent truly new and original ideas to get a foothold and develop. The information race smoothes out the novelities and irregularities of the landscape of ideas. Obviously this factor is more influential in some areas than in others. In the commercial world there are ways and means (such as patents) to protect crucial new developments. In other areas such as politics every statement and action is met immediately with feedback responses from the media and the public. Is it possible to foster creativity under such circumstances?

It would be rash to infer that isolation is a necessary condition for all creative work. There are many variations to this theme and it might be of interest to see how the creative process differs under different circumstances with respect to interaction and external stimulii. In science where the outcomes (more easily than in the arts) can be measured against some accepted standards or reference, the lack of external contacts can lead to rather curious results. A famous example is the Indian mathematician Srinvasa Ramanujan. He was born in the 1880s in the Tanjore district of Madras. He had an extraordinarily mathematical intuition and made mathematical discoveries and rediscoveries without any real contact with the mathematical traditions. He was "adopted" by G.H. Hardy and came to Cambridge for a few years before he contracted tuberculosis and died at an age of thirty-three. Ramanujan had no interest what so ever to relate his mathematics to anything. He was interested solely in numbers and their relations. A quote from Hardy's biographer Robert Carmichael (cited in John Barrow's book Pi in the Sky): "In some directions his knowledge was profound. In others his limitations were quite startling ... he had conceived for himself and had treated in an astonishing way problems to which for a hundred of years some of the finest intellects in Europe had given their attention without having reached a complete solution. [...] What is astonishing is that it ever occurred to him to treat these problems at all".

CREATIVE ENVIRONMENTS

The dream of creating the ideal conditions for creative scientific work has been probably found its best modern expression in the Institute for Advanced Study at Princeton, founded in 1933 as a place dedicated to new and fundamental explorations. No students, no teaching and no worldly worries, a place for the most able intellectuals to think and write. No laboratories and experimental work. A truly Platonistic heaven. Princeton did became a famous institute with the best mathematicians and physicists around, hosting a dozen or more Nobel prize laurates. The first and most famous professor was Albert Einstein. In spite of the success story of Princeton (it is still a world leading institution especially in mathematics) the sceptical asks the question: Does such an intellectual meltingpot as Princeton Institute provide optimal conditions for creative scientific work? One of the critical commentaries, the physicist Richard Feynman, made the following assessment of the institute where all these great minds were gathered: "Nothing happens because there's not enough real activity and challenge. You're not in contact with the experimental guys. You don't have to think how to answer questions from the students. Nothing!"

It is a fact that Einstein did not make any major achievements during his more than twenty years at Princeton. He was working on grand unified theory trying desperately to get around quantum mechanics and the Copenhagen interpretation but with no success. (The possible exception being the gedanken experiment known as the Rosen-Einstein-Podolsky paradox which was formulated in 1935 two years after that Einstein arrived to the institute.) Another example is Kurt Gödel who also came to Princeton in 1933 and remained there until his death 1978. He was already world famous for his incompleteness results (proving the falsity of Hilbert's conjecture about the completeness and consistency of arithmetic). Gödel spent his time trying to solve the continuum problem. He made some contributions but failed to reach his goal (the continuum hypothesis is still an open problem).

A common objection to this argument at least in the natural sciences and in mathematics would be that old people are getting less creative anyway regardless if they are at Princeton or not. Nobel prize winners seldom make any significant discoveries after they received the prize. It might well be true that some of the intellectual ability to make truly original contributions in logically demanding subjects such as mathematics declines with age. (Counter to general beliefs there seems to be little emprical evidence however that productivity in mathematics diminishes with age!) Another explanation which might be equally relevant (mentioned by the mathematician René Thom) is that senior scientists simply are not interested in the same kind of problems. You tend to shift your attention from virituoso problem solving to broader and maybe more basic problems which present a different sort of challenge and where solutions are not so clearcut. You become more philosophical when you get older.

COLLECTIVE CREATIVITY

Creativity is not only associated with individual ability but can also be seen as a collective property resulting from interaction and collective processes. Brainstorming is a well known expression for a form of collective creativity.

In science this would be reflected in joint work and co-authorship of articles. In empirical and experimental oriented sciences the sheer amount of information can be overwhelming (there are such vast amounts of data around—untouched by human thoughts!) and the technical work itself demands a great deal of cooperation efforts. Big science, such as high energy physics, would not be possible without the input from many people. If you browse through scientific journals today you find a great deal of joint papers and that portion is increasing. In some fields you might typically find a dozen or so authors and co-authors.

The style of science is changing. But does co-authorship reflect trutly cooperative creativity? Does collective creativity have similar qualities as individual creativity or what are the differences? There are certain kinds of problems which are opportune for joint work and where there is a definite added value to be gained from interaction. Crossword puzzles is a good metaphore to illustrate this difference. You can benefit a great deal from being a group trying to solve the puzzle together. But constructing a crossword is typically a one person job! In the latter case the demand for "inner consistency" is probably so great that it is less efficient to try to communicate this quality while you are trying to put the whole thing together. The same obviously holds for most art. To write a symphony you probably want to be alone, but to play it you call on your friends.

So we would make the hypothesis that collective creative work by its own logic represents a different perspective and finds other expressions than problems formulated and solved by single individuals. If these differences are significant in how problems are perceived and formulated it would be of relevance in the discussion about interdisciplinary science. Today a lot of scientific programs are formulated as interdisciplinary endevours where representatives from different scientific backgrounds come together and contribute to a common area of study. The fragmentation of scientific knowledge has made such an approach almost mandatory. It resembles the crossword solving exercise. But who constructs the crossword? To define the problem in the right way is probably the most critical part in any scientific process. As Steven Weinberg points out in his book Dreams of a final theory: "The historical importance of quantum mechanics lies not so much in the fact that it provides answers to a number of questions about the nature of matter—much more important is that it changed our idea of the questions that we are allowed to ask". To envisage the right connections and to exclude and realise what is irrelevant. Someone has to have the vision and dare to make the leap.

ASSOCIATIVE AND LATERAL THINKING

The ability for associative thinking is probably a decisive factor in creativity. Again Newton can serve as an example. In his case we can talk about association on a paradigmatic level. He surely did not make progress solely by thinking hard on problems. He also calculated and made experiments. "No other investigation of the seventeenth century better reveals the power of experimental enquiry animated by a powerful imagination and controlled by rigorous logic", writes Richard Westfall in his Newton biography. The scientific method had proven its tremendous power. The scientist is not alone in his task of combining imagination with observation. A creative artist would also explore this duality as Igor Stravinskij once remarked: "We never get the ability to create alone, it always come together with the ability to observe".

Many creative acts are the result of surprising combinations of knowledge from different domains. Mathematics is full of such powerful interconnections, on a general level such as between algebra and geometry but of course also on very specific issues as the 1994 Crafoord prize in mathematics illustrates. It was awarded to Simon Donaldson for his work of instantons and the geometry of fourmanifolds, two highly specialised areas seemingly far apart. Typically the only question raised after his prize lecture was this: How did you come to think about this connection?

RULES AND CONSTRAINTS

As we already have pointed out, the success of Newton's research was not only due to his sharp thinking but also to his well controlled experimental work. Unlike many of his predecessors he did not satisfy himself with speculation and he did not start out trying to create an all embracing natural philosophy. He was very specific and he constrained himself to phenomena which he could observe and measure with great precision. He was creative under rigorous constraints. The importance of constraints has been witnessed by many. It is a theme that also runs through the discussion about creative art. Goethe said: "In der Beschränkung zeigt sich erst der Meister". Stravinskij expresses a similar opinion when he confesses that: "My freedom growths and deepens as I diminish my field of activity and puts up more obstacles in my way". He sees the role of the creative human to sift the elements she receives from imagination and mark out boundaries. "The more art is controlled, confined, cultivated the more free it is". Play and art without rules is uninteresting. Absolute liberty is boring. As the Swedish writer Lars Gustafsson pointed out: "It is like playing tennis without a net".

Creation in science is very much a matter of interaction with rules and constraints. Much of scientific theories get their sense of importance from obeying rules, which make theories inevitable. One of the most powerful type of principles which makes modern physics so elegant and convincing can be described as principles of symmetry, i.e. the postulate that the natural laws are invariant under certain transformations. These apply not only to space and time as in classical theory or to all frames in motion (such as in general relativity implying the existence of gravitation) but they are applicable for viewing type and identity of elementary particles (internal symmetries).

Principles of symmetries gives a kind of rigidity to theories imposing restrictions on which particles and forces could and should exist.

The fact that we see symmetry in the laws and equations but not necessarily in the individual solutions gives an interesting twist to the difference between the elegant world of physics and the complex world of biology and the social sciences, sciences which deal with the results of a long evolutionary process—a process of symmetry breaking realisations of the physical laws.

As important as rules and conventions are for building and maintaining a common mass of knowledge equally critical is the rule breaking and transcendence of conventions for creating new things. Creativity finds is fertile soil in the conventions. Mastery reveals itself as breaking rules. The secret of creativity hinges on this insight, to know the right moment when you can go too far. Certainly such an insight must be related to a sensitivity and openess, an openess to the accidental, to chance, to noise, and an ability to listen to weak signals. Milan Kundera touches upon this idea in a more poetic form in his novel *The Unbearable Lightness of Being*: "Only chance can be interpreted as a message. That which happens by neccessity, that which is expected and is repeated daily is mute. It is only chance that speaks to us".

UNCERTAINTY AND AMBIGUITY

To create something new implies stepping out of the ordinary and familiar. It is a marked difference in being original and being creative. Creativity has a special quality of mental fecundity passed through a sieve of judgement. It is not just a willful choosing of the uncommon. Uncertainty and ambivalence play a crucial role in the initial stage. It is perhaps no accident that a creative process often starts with the scribbling down your ideas on a napkin or on the table cloth. The napkin is not only a practical device but it is also a representation of the fuzziness and playfulness of the situation! (Being on the edge of chaos, to use a popular metaphor.)

It is part of our conception of creativity to see a creative person as a person with great sensitivity, exposing herself to external impressions that ordinary people would shun away from. Also with more open channels between the conscious surface and the subconscious depth. Creativity involves both intoxication and anxiety. There are many examples to support this view. One such case is the Swedish author August Strindberg. He has been read, studied and analysed in every conceivable way. In a recent book the Swedish psychiatrist Peter Curman has analysed Strindberg's so called inferno crisis and demonstrated how he manage to transform his emergent insanity into productive work such as *The Dream Play*.

An interesting aspect in this context is that Strindberg in spite of his chaotic inner life was a very rigid and pedantic person. The same holds for Newton. Although no comparisons should be made in terms of mental instability Newton also seems to have been very rigid in his habits. He was meticulous, well organised and good at administration. He reorganised and ran the British Mint for many years and he became the president of The Royal Society. Both these jobs required a great deal of management skill. A curious detail about Newton's well organised and inquisite mind is that he is reported never to read attentatively without a pen in his hand. And he copied a lot. It has been speculated that his excessive copying stemmed from a conviction that he could rely with confidence only on himself alone.

FINAL REMARKS

Where do we find the extrordinary creativity in today's society? Naturally the legacy of Newton and Einstein is carried forward in the scientific community. There is no reason to believe that creative achievements in science have come to an end. Still the daily work of the ordinary researcher in the laboratory is far from pathbreaking revolutionary events we associate with the scientific revolution in the 17th century, Darwin's theories in the 19th century or the creation of relativity theory and quantum mechanics in our own century.

The interest for creativity is however greater than ever. We seem to be looking for agents for change that can help us to exploit the dynamics and uncertainty in contemporary society in beneficial and productive ways. It is by no means obvious that the academic institutions will be leading the way in driving this development forward. On the contrary, the university have had a traditional role of being a critical basis for society and to work against the prevailing spirit of the time. This spirit of today is very much dominated by rapid change and short term market forces dominating the scene. My guess is that it is in this stream of activities we would tend to identify the present sources of extraordinary creativity rather than in the secluded academic environments such as Trinity or Princeton. If this article were to be written again in 50 years time and the author would be looking for examples from our time I think that people like Steven Spielberg (with Jurrasic Park) or Bill Gates (with Microsoft) would come to mind more easily than our university professors and academy members.

Swedish Polar Research Secretariat and
Royal Institute of Technology, Sweden

REFERENCES

Barrow, J., *Pi in the sky*, Clarendon Press, 1992.

Curman, P., Skaparkriser – Strindbergs Inferno och Dagermans (Crisis of Creativity), Natur och Kultur, 1992.

Stravinskij, I., *Poetics of music in the form of six lessons*, Harvard University Press, 1970.

Sörlin, S., *De lärdas sällskap* (The Learned Society), Liber, 1994.

Regis, E., *Who got Einstein's office?*, Penguin books, 1987.

Weinberg, S., *Dreams of a final theory*, Vintage, 1993.
Westfall, R., *The life of Isaac Newton*, Cambridge University Press, 1993.

JOHN L. CASTI

THE WORLD, THE MIND AND MATHEMATICS

IS APPLIED MATHEMATICS BAD MATHEMATICS?

Mathematics is a field full of dichotomies: continuous versus discrete, size versus shape, finite versus infinite and static versus dynamic to name just a few. But certainly the most controversial division is the notorious "pure" versus "applied". And in a provocative 1981 article [1] titled "Applied Mathematics is Bad Mathematics", Paul Halmos set the fox in among the chickens by claiming that usually the answer is Yes!, and that the reason ultimately comes down to a question of taste. To use one of Halmos's analogies, a Picasso portrait is usually regarded as better art than a police photograph of a wanted criminal. But the Picasso is probably not a very good likeness, while the police photo is far from inspiring to look at. So is it completely unfair to say that the portrait is a bad copy of nature and the photograph is bad art?

This is the crux of the argument in support of pure mathematics as being somehow a higher art form than the applied. Well, maybe it is, although many minds far more perceptive than mine have put forth good arguments against it. And it's not really our intention here to jump into this seemingly interminable debate. Rather, we want to explore the dimensions of a more recent, and somewhat less well-known mathematical dichotomy, the distinction that one sometimes sees today made between "applied" and "motivated" mathematics. So before stating our purpose, let's first set out the terms of the discussion to follow.

THE ONTOLOGY OF MATHEMATICS

A debate that has been raging with equal vigor — and probably a lot longer — than that over pure versus applied mathematics is the one over the nature of mathematical objects. Are things like a right triangle, a semisimple Lie algebra or the number π *real* objects like, say, Mount Everest? Or are they pure inventions of the human mind? Mathematicians holding to the bona fide reality of such objects living in some realm beyond everyday space and time, have been labeled "Platonists" for obvious reasons, since their brand

This paper was prepared at the Institute for Future Studies, Stockholm, Sweden, whose support is gratefully acknowledged.

Å. E. Andersson and N.-E. Sahlin (eds.), The Complexity of Creativity, 115–138.

of mathematical reality is a reflection of mathematical objects living among Plato's famous Ideal Forms. For those who feel that these types of mathematical gadgets are pure inventions of the human mind, such Platonic leanings seem at best nonsensical. Mathematicians are pure inventors, they argue, no different in kind than Thomas Edison or Giuseppe Marconi, other than the irrelevant fact that the products of the mathematician's invention are carved from pure form, structure and information rather than from tangible matter.

So are mathematicians explorers, discovering imperfect manifestations of eternal Platonic forms? Or are they inventors, creating the concepts and structures as needed out of whole cloth? The position one takes on this issue reflects perfectly the distinction between "applied" and "motivated" mathematics.

For our purposes, an *applied* mathematician is a discoverer. Such a person faces a problem from the world of nature or mankind, and attempts to solve the problem by developing a mathematical picture of the situation. The applied mathematician paints this picture by taking existing mathematical notions and structures and putting them together in new ways. Of course, it often turns out to be the case that all of what's needed for the picture is not available in "off-the-shelf" form, so new "colors" and/or "brushes" have to be created to paint the picture. But the basic focus of the applied mathematician is on solving the real-world problem, and the creation of new mathematical tools is secondary to this task. So it is in this sense that he is a discoverer; he discovers how to arrange existing objects to accomplish a given task.

The development of the theory of probability by Pascal and Fermat is a good example of applied mathematics. It's doubtful that either Pascal or Fermat had in mind the development of a new branch of mathematics in 1654, when they initiated their attempt to answer the Chevalier de Meré's question about the division of stakes in a dice-throwing game. Nevertheless, the intellectual roots of the modern theory of probability were laid down as a consequence of grappling with that *very* real problem from the world of gambling.

The *motivated* mathematician, on the other hand, gives primacy of purpose to the creation of new mathematics. The real-world serves mostly as a source of inspiration to such a mathematician, not principally as a source of work. It was the motivated mathematician that John von Neumann had in mind when he wrote in his essay "The Mathematician" [2] that "mathematical ideas originate in empirics".

To illustrate motivated mathematics in action, consider the famous Three-Body Problem of celestial mechanics. In 1887, King Oscar II offered a prize of 2,500 crowns for an answer to the question: Is the solar system stable? The prize was finally awarded to the great French mathematician Henri Poincaré in 1890 for his monumental treatise *On the Problem of Three Bod-*

ies and the Equations of Dynamics. So what started out as a very definite problem about the behavior of planetary bodies was transmogrified into a mathematical problem about an *idealized* "solar system", one in which there are only three gravitating bodies, each regarded as a point mass moving in a frictionless medium. Once formulated in this purely mathematical fashion, the original celestial motivation faded away like a trickle of water in the desert. But the physicist's loss was the mathematician's gain, as the Three-Body Problem has spawned the entire field of dynamical systems, not to mention much of modern topology and geometry.

We will cite several more contemporary examples of both applied and motivated mathematics later on. For now, let's briefly state our objectives. In the remainder of the paper, we want to address the following two central questions:

- Is there any distinction that matters, mathematically or otherwise, between these two views of mathematics and its relation to the real world?
- If there is a perceptible difference, can one rank order the two types of mathematics on the basis of the creative content of their respective products?

Before jumping into these questions with both feet, let's first talk a bit about creativity, first in general, then more specifically in the case of mathematics.

"BIG C" CREATIVITY

From the point of view of scientific analysis, the term "creativity" is a slippery one. This is mostly due to the fact that all of us think of ourselves as being creative, and so almost every general statement made about creativity immediately runs up against the personal experience of the reader. So at the beginning, let's agree to distinguish between the sort of creativity that permeates our daily lives, like combining a red sock with a blue one — "little C creativity" — and the kind of conceptual leaps of the intellect that occur only seldom — "big C creativity".

In [3], the cognitive psychologist Howard Gardner has advanced a useful definition of this second type of creativity. Gardner writes, "a creative individual is one who regularly solves problems, fashions products and / or poses new questions in a domain in a way which is initially considered novel but which is ultimately accepted in at least one cultural setting". Let's tease apart the components of this definition.

- *Problem-solving*: Whatever creativity may mean, virtually every investigator agrees that it involves solving some sort of problem. In fact, most definitions restrict creativity to just this activity alone.

- *Products*: Very often, however, the output of a creative act is not a problem solved but a product produced — a sculpture, a scientific theory, a musical work. Of course, these products themselves are often the solution to a problem. But it is the product that lives on, not the problem that gave rise to it.
- *Novelty*: As with problem-solving, almost all studies of creativity note that creative acts are initially original but that they eventually become accepted. Moreover, if they are not accepted, then they are not considered creative. Weird, unusual, anomalous perhaps, but not creative.
- *Cultural acceptance*: A creative act by the above definition is like the proverbial tree falling in the forest: it cannot be creative unless it is recognized as a creative act by someone besides its creator. In particular, creativity is not creativity unless it is deemed so by a relevant social institution. Such an institution may be nothing more than a small group of knowledgeable peers. Or it may be an entire establishment, as in literature, consisting of other writers, critics, editors and the reading public, at large.

The above definition of creativity opens up a question first posed by psychologist Mihalyi Csikszentmihalyi, who asked in [4]: "Where is creativity?" So instead of focusing exclusively on the individual by asking "Who is creative?" or even "What is creativity?", both this question and the definition just given suggest that creativity does not reside at a single location but rather in the interaction among several locations. Specifically, we can identify three nodes that taken together prescribe the locus for creativity. These are: (1) the *individual person*, (2) the formal structure of knowledge in an *area* or domain, and (3) the institutional structure, or *field*.

As an illustration in mathematics of this triadic view of creativity, consider the field of chaotic dynamics. There are currently hundreds of researchers and students around the world working on developing this domain of dynamical system theory. Each of them must present proofs of theorems and other results in the domain, which in turn are addressed to other workers in the field. In the natural course of things, only a few of the products of these workers will actually stand out strongly enough in the domain of chaotic dynamics that the work will be encountered as "classical" by the next generation of dynamical system theorists. So here we find the individual researcher working in a given domain — chaotic dynamics — whose work is judged creative or not by an institutional mechanism, the field, consisting of the people running the mathematical journals and academic departments.

With this example in mind, it's instructive to look at each of these nodes in a bit more detail.

- *The Individual* — There are two aspects of creativity as it pertains to the individual creator. The first is centered on the relationships between the various types of intelligence possessed by the individual. For example, the logical and mathematical abilities of an Einstein or the gift for spatial arrangement of a Leonardo. In short, the creative act usually emerges from an individual having an "excess" amount of one kind of intelligence, often at the expense of a "deficiency" in other types.

The second aspect of creativity that comes into play at the individual level is the kind of intimate relationships that the person has at the time of a creative breakthrough. While creative people tend to be fiercely independent and are often loners, it seems that at the time of their breakthroughs they need to have an intimate to whom they can confide their thoughts. Examples include T.S. Eliot with Ezra Pound and Albert Einstein with Michel Besso.

- *The Domain* — As with the individual, there are also two aspects of creativity associated with the area of activity. The first has to do with how highly structured the domain is. Areas like mathematics and academic literature are rather well-structured, in the sense that there are many well-defined levels that one passes through on the way to expertise. Other areas like modern art and the development of computer software are much less well-structured.

  A related aspect is the extent to which there is a dominant paradigm within the problem area, as opposed to several competing paradigmatic structures. The current field of complexity studies is a good example in which there is currently no dominant paradigm, as a number of theories of complexity are competing for attention and recognition. On the other hand, evolutionary theory has a very well-defined structure, the neo-darwinian paradigm, within which virtually all workers state and evaluate their ideas.

- *The Field* — At this node, too, there are two important aspects. One involves the size of the field, and especially its hierarchical ordering. In Einstein's time, the number of physicists was small, and only the opinion of a handful of them — Lorentz, Planck, Bohr — mattered in assessing work for its novelty and contribution to the field. The second aspect of the field that is important is the degree of consensus. In some periods, there may be a wide consensus as to what constitutes the important questions, methods and criteria; at other times, it's just the opposite, with great divisions separating the field.

Figure 1 shows an overview of this triadic breakdown of the structure of creativity.

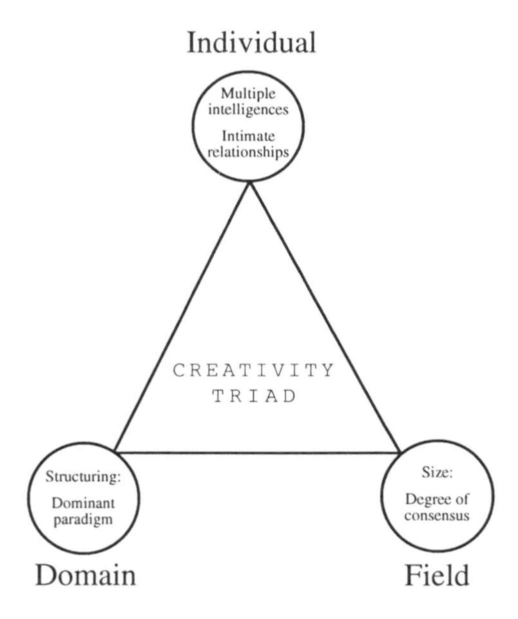

Figure 1. *The triadic structure of creativity.*

On-going work by Howard Gardner and others suggests that creativity is particularly likely to occur when there is some degree of tension within or across these three nodes. For example, there may well be some internodal tension between competing types of intelligences at the individual node or between competing paradigms at the level of the domain. But the real creative sparks start to fly when tension arises across nodes.

For example, early in this century classical music went off into two very different directions: the rhythmic, polytonal music of Stravinsky, and the atonal, twelve-tone music of Schoenberg. This was a tension between the domain and the field. And at the time it was far from clear which of these directions would ultimately win out. It now appears that twelve-tone music has fallen by the wayside, and that the field has voted with its ear, so to speak, against serial music and in favor of Stravinsky.

With these general notions of creativity in hand, let's turn to our primary focus, considering some of the ideas that have been put forward for the specific case of creativity in mathematics.

MATHEMATICAL CREATIVITY

Certainly the most well-known treatment of creativity in the mathematical field is the celebrated lecture "Mathematical Creation" [5] (also known in some translations as "Mathematical Discovery") by Henri Poincaré, which is the text of an address given to the Société de Psychologie in Paris. By Poincaré's account, the mathematician is more like an artist than a scientist in his approach to his task, looking for patterns and structures by a combination of intuition, judicious selection from an infinitude of possibilities and an overdose of that ineffable quality we generally refer to as "taste".

Of special interest in Poincaré's treatment is the high premium he places on the work of the unconscious mind. Poincaré gives a well-known autobiographical example of how after weeks of grappling with a problem involving fuchsian theta functions, he set the problem aside only to have the complete solution appear to him, almost as if by magic, as he was stepping onto a bus. It appears that his unconscious mind continued to grind away at the problem, even after his conscious mind had dropped it in favor of other concerns. Poincaré argues that this period of feverish conscious work, followed by an interval during which the problem lies fallow, and then the solution appearing in a flash is characteristic of creative work in mathematics. Let's consider another case.

In the essay "The Mathematician", referred to briefly above, John von Neumann presents a convincing case that much of the best mathematical inspiration comes from experience, but that mathematics itself is very far from being an empirical science. Thus, it is often the aesthetical character of a particular piece of work that determines its merit, not its usefulness. This echoes remarks made by the British mathematician G.H. Hardy, who in [6] took the extreme view that the value of a piece of mathematics is inversely proportional to its utility.

Von Neumann goes on to list several additional features of creative genius in the product of outstanding mathematical work — simplicity, elegance, depth — all properties that have been remarked upon regularly by all writers on the topic.

More contemporary writers on the theme of creativity in mathematics have also noted many of the same features. For example, Paul Halmos writes [7]:

> For the professional pure mathematician, mathematics is the logical dovetailing of a carefully selected sparse set of assumptions with their surprising conclusion via a conceptually elegant proof. *Simplicity, intricacy, and above all, logical analysis are the hallmark of mathematics.* (italics added)

Halmos goes on to note that "mathematics is never deductive in its creation", closing his essay with the remark that "perhaps the closest analogy is between mathematics and painting . . . Almost every aspect of the life and of the art of a mathematician has its counterpart in painting'.'

From the triadic perspective, what is most striking about these accounts of the creative process in mathematics (and the list could have been extended considerably) is the total preoccupation with the node labeled "Individual" in Figure 1. The focus of all these accounts is on what the *individual* mathematician does, thinks and feels during the course of creating beautiful new mathematics. Pushing such accounts to their extreme, one might be led to believe that great mathematics exists in a vacuum, independent of either the *Zeitgeist* of the area or an institutional gatekeeping mechanism. When put in such bald fashion, I doubt that even G.H. Hardy would defend such a view. Nevertheless, the descriptions given by Poincaré, Hardy, Halmos and others of creativity in mathematics leave the impression that great work can be done in absence of any external influences. Neither what's "hot" in a particular area of mathematics, nor what the mathematical establishment think of the work play any role whatsoever in these descriptions of creativity.

With this point in mind, let's return to the question that started this essay, the interplay between creativity and the real or imagined dichotomy between applied and motivated mathematics. First, we look at a modern example from economics showing some of the character of both applied and motivated mathematics. Then we'll look at two other examples, also from economics, that seem to be quite clearly in one camp or the other.

EQUILIBRIUM ECONOMICS

As envisioned by Leon Walras in his path-breaking 1874 book *Eléments d'Economie Politique Pure*, the observed state of an economy can be regarded as an equilibrium resulting from the interaction of a large number of agents with partially conflicting interests. But as soon as an equilibrium state is defined for an economy, the question arises as to its existence.

So successful was Walras's mixture of mathematics and economics, that a century later this line of investigation received the ultimate legitimization when Kenneth Arrow was awarded the 1977 Nobel prize in economics, mostly for work on what has come to be termed "general equilibrium theory". And to add a bit of frosting to the cake, the Nobel committee honored Gerard Debreu in 1983 for further developments of general equilibrium theory reported in his magnum opus *Theory of Value*, a volume of such dazzling mathematical pyrotechnics that it turns completely upside down the traditional view of mathematicians as being the keepers of the abstract processes, that dubious honor having now clearly passed on to the economists. Here is a simple example of equilibrium economics, showing how the ideas of supply, demand and prices relate to the existence of fixed points of a transformation.

In Walras's view of economic man (and woman), the agents in the economy are divided into two classes: consumers, who wish to make use of the goods and services of the economy, and producers, who transform what the consumers own into other goods that consumers desire. Thus, the stocks of commodities in the economy are all assumed to be owned by consumers — either in their tangible form or by a variety of financial instruments such as stocks in corporations and government bonds.

If the prices of all goods and services in the economy are known, the wealth of each consumer is determined by the market value of his or her assets. Therefore, income and a knowledge of prices permit a consumer to state demands for commodities, as well as what he or she is prepared to offer by way of labor, raw materials, money and other commodities that then become available to the producers. In the model outlined below, these market demands will be a function of the relative prices of all goods and services.

As an elementary, prototypical formalization of this Walrasian vision of the supply-demand equilibrium problem, think of an exchange economy E with l commodities and a finite set A of consumers. We let x_a represent the consumption vector of consumer a. This is a vector in the space R^l_+, whose elements are just non-negative real numbers. Thus, the ith component of this consumption vector is just the amount of the ith commodity consumed by a.

Next, we define a price system in the following way. Let p be a vector in the interior of R^l_+. The ith coordinate of p is the amount paid for one unit of commodity i. So given the price vector p and consumer a's wealth vector $w_a \in R_+$, this consumer must obey the budget constraint $p \cdot x_a \leq w_a$. Since scaling prices and wealth by a positive number has no effect on consumer behavior, it's convenient to normalize the price vector p so that it lies on the surface of the unit sphere in R^l, i.e., $\|p\| = 1$.

Given the wealth w_a, we postulate that consumer a has a demand given by the continuous consumption function $f_a(p, w_a)$, which is also a vector in R^l_+. And to complete the description of the economy E, we specify an initial endowment vector e_a for consumer a. This just represents the particular bundle of commodities actually in the hands of consumer a. Thus, for a given price vector p, the wealth of consumer a is $w_a = p \cdot e_a$. So the characteristics of consumer a are given by the consumption function f_a and the endowment vector e_a. The entire economy E is then represented by the list of pairs $\{(f_a, e_a): a \in A\}$.

If consumer a's demand is given by the consumption function $f_a(p, w_a)$, then the excess demand in the economy is

$$F(p) = \sum_{a \in A} [f_a(p, w_a) - e_a].$$

The price vector p is then an equilibrium price vector if and only if this excess demand is zero, i.e., $F(p) = 0$. If we assume that every consumer is insatiable, in the sense that he spends his entire wealth, then we have the spending condition that $p \cdot f_a(p, w_a) = 0$. This leads to what is termed *Walras's Law*: $p \cdot F(p) = 0$. It's clear that F is a continuous vector field on the set of prices P. The task is to show that there exists a price equilibrium vector $p^* \in P$, since only if such a set of prices exist can there be a state of the economy in which all consumers are satisfied.

In order to establish the existence of a set of equilibrium prices, we make the following assumption about the behavior of the excess demand function

F near the boundary of the price set P (where, by definition, at least one of the commodity prices is zero): If p_n tends to a point p_0 on the boundary of the price set, then the sequence $\{F(p_n)\}$ is unbounded. In plain English, this assumption says that every commodity is collectively desired. With this in mind, we have the

EQUILIBRIUM PRICE THEOREM. *If the excess demand function F is continuous, bounded from below, and satisfies Walras's Law and the collective desirability condition, then there exists an equilibrium price vector* $p^* \in P$.

So here we have what looks like an example of applied mathematics, *par excellence*. Start with a real economy, formalize (i.e., abstract) the economy in mathematical terms, and then take whatever existing mathematics you need in order to answer questions about the structure of this mathematical world. Of course, it may well be necessary to extend or generalize the existing mathematics to fit the situation. And, in fact, that was exactly the case with equilibrium economics, where people like Arrow and Debreu developed sophisticated extensions of original work by Brouwer, Schauder and Kakutani to *create* fixed-point results tailor-made for economic questions. But the applied versus motivated character of these results becomes clouded when we look a bit deeper into the history of both Arrow's and Debreu's work.

Arrow came at the question from a purely applied mathematical point of view. As he himself notes in [8]:

> My work in economics took its original cast from the depression ... My ideal in those days was the development of economic planning, a task which I saw as synthesizing economic equilibrium theory, statistical methods, and criteria for social decision making. ... Naturally my concepts and directions were partially altered by my own development and by changes in the world and in economic science as a whole.

Gerard Debreu's work, on the other hand, seems to start from just the opposite direction. Although his subject matter is an exchange economy, and his ultimate goal is still to give an account of the formation of exchange values, Debreu insists that the theory's formal structure must be constructed axiomatically and with no reference to the interpretive values of the concepts. In his pioneering work [9], he states:

> Allegiance to rigor dictates the axiomatic form of the analysis where the theory, in the strict sense, is logically entirely disconnected from its interpretations. In order to bring out fully this disconnectedness, all the definitions, all the hypotheses and the main results of the theory, in the strict sense, are distinguished by italics; moreover, the transition from the informal discussion of interpretations to the formal construction of the theory is often marked by one of the expressions: "in the language of the theory", "for the sake of the theory", "formally". Such a dichotomy makes possible immediate extensions of that analysis without modifications of the theory by simple reinterpretations of concepts.

So general equilibrium theory is one area of economics in which both applied and motivated mathematics have given rise to Nobel glory. Now what about an example of motivated mathematics, pure and simple?

IT'S ALL IN THE GAME

Think of the dilemma faced by Peter and Reneé, two young entrepreneurs, who plan to locate a new restaurant at a major intersection in the nearby mountains. They agree on all aspects except one: Reneé likes low elevations, while Peter wants to go as high as possible.

The dimensions of their location problem are laid out in Table 1, where we see that there are three routes, Avenue A, Route B and Road C, all of which run in the north-south direction. Furthermore, there are also three highways, numbered 1, 2 and 3, which run east and west. The table gives the altitude in thousands of feet at the intersection of these thoroughfares.

	Highways		
Routes	1	2	3
A	10	4	6
B	6	5	9
C	2	3	7

Table 1. *Altitudes at road intersections*

Peter and Reneé agree to make their decision in the following way: Peter will select one of the three routes A, B or C, and Reneé will simultaneously pick one of the three highways 1, 2 or 3. The restaurant will then be located at the intersection of these two choices.

If Peter is pessimistic and thus does a worst-case analysis of the his situation, he looks at the lowest elevations of the three routes (the minimum of each row of Table 1) and then chooses the largest of these three numbers. If Reneé is similarly pessimistic, her analysis involves examining the highest of the three highways (the maximum of each column of the Table) and then selecting the lowest of these numbers. Carrying out these operations on Table 1, we see that these two numbers, Peter's maxmin and Reneé's minmax coincide at the entry Route B, Highway 2. This is the so-called *minimax solution*, or the equilibrium point of the game between Peter and Reneé. Thus, the optimal strategy for Peter is to choose Route B, while Reneé's best choice is Highway 2. Moreover, there is no need for secrecy in this game, since even if Peter were to reveal his choice of Route B in advance, Reneé would not be able to use this knowledge to her benefit.

The minimax solution of the above game is an example of what in the theory of games is called a *saddle point*. And it's easy to show that in two-person games in which one player's gain is the other's loss, the saddle point is the optimal strategy for both players. This will be the case for those payoff matrices like Table 1 where the maximum of the row minima equals the minimum of the column maxima. But only very special payoff structures will have this property; hence, not all two-person games have saddle points. And in such cases, if one player announces his or her action in advance, the other player can exploit this information.

One way out of this dilemma is for the players to mix their actions, using perhaps a randomizing device like the flip of a coin to decide what action to take, so that even the players themselves do not know in advance what they will do. Such a probabilistic course of action is called a *it mixed strategy*. In these cases, the payoffs to the players can only be measured in stochastic terms by the *expected returns* they would receive if the game were played many times. But then the question arises: Do there exist optimal mixed strategies for each player? Here, of course, "optimal" means in the sense that the expected minimax payoffs to the two players is the same. Let's see.

Suppose the components of the vector $p = (p_1, p_2, \ldots, p_m)$ represent the probability that Player I chooses action i, $i = 1, 2, \ldots, m$. We call p the *strategy vector* for Player I. Similarly, the vector $q = (q_1, q_2, \ldots, q_n)$ is the strategy vector for Player II. Further, let r_{ij} be the payoff to Player I when he selects action i and Player II selects action j. Suppose the game is zero-sum. Then the payoff to Player II is simply $-r_{ij}$, and the *expected* return to Player I is just the sum

$$\sum_{i=1}^{m} \sum_{j=1}^{n} p_i q_j r_{ij}$$

Player I clearly wants to choose his strategy vector p so as to maximize this quantity. Similarly, Player II tries to select the strategy vector q to minimize this sum. The question of the moment is to ask if there is a choice of strategy vectors for the two players that will result in

$$\max_p \min_q \sum_{i=1}^{m} \sum_{j=1}^{n} p_i q_j r_{ij} \min_q \max_p \sum_{i=1}^{m} \sum_{j=1}^{n} p_i q_j r_{ij}$$

In 1928, following up on work by Emile Borel who analyzed this question in the special case when $m, n \leq 4$ and Ernst Zermelo who conjectured that equality should hold in the above relation, John von Neumann proved the famous

MINIMAX THEOREM. *For a two-person, zero-sum game with payoff matrix $R=[r_{ij}]$, there exists a unique number*

$$\max_p \min_q \sum_{i=1}^{m} \sum_{j=1}^{n} p_i q_j r_{ij} = V = \min_q \max_p \sum_{i=1}^{m} \sum_{j=1}^{n} p_i q_j r_{ij}$$

and strategy vectors p^* and q^* such that

$$\min_q \sum_{i=1}^{m} p_i^* q_j r_{ij} = V = \max_p \sum_{j=1}^{n} p_i q_j^* r_{ij}$$

Here the quantity $V=V(p^*, q^*)$ is termed the value *of the game.*

So just as the introduction of complex numbers restored the solvability of any polynomial equation, Von Neumann's Minimax Theorem restores the solvability of any two-person, zero-sum game by guaranteeing the existence of a saddle-point equilibrium — but now in the space of mixed rather than pure strategies. By extending the notion of what we mean by a strategy from the choice of a single course of action to a randomization over all possible actions, von Neumann succeeded in establishing the existence of a rational choice that either player can announce in advance without giving the opponent any sort of advantage. It's no wonder that von Neumann could later remark that, "As far as I can see, there could be no theory of games . . . without that theorem . . . I thought there was nothing worth publishing until the 'Minimax Theorem' was proved".

Historically, game theory as developed by von Neumann and the economist Oskar Morgenstern, was motivated by the hope of creating a rational theory of economic processes. In their pioneering 1944 work, *Theory of Games and Economic Behavior*, von Neumann and Morgenstern tackled a variety of problems motivated by situations in economics, focusing special attention on those in which there are more than two players.

Subsequent work by a virtual army of researchers has developed the theory of games into an imposing mathematical enterprise, whose roots in economic processes is barely noticeable. So here we have a perfect example of motivated mathematics, a theory that started its life as a mathematical response to problems in an already idealized economy, and which went on to become a branch of pure mathematics. Now to close our account of modern mathematics, both applied and motivated, let's look at an example of purely applied mathematics.

UP AT THE CORNER

Many management decisions ultimately come down to decisions about how to allocate resources so as to optimize something, e.g., the allocation of money to maximize return on investment or the allocation of people and materials to minimize the total cost of producing a product like a car or a TV set. Often, these kinds of problems can be formulated in such a way that they can be solved by what's called "linear programming". Before describing this idea in general terms, let's look at a very simple example just to get the general idea.

Consider the situation faced by the Chow-Down Dogfood Company, which manufactures two types of dog food, the Bow-Wow and the Wuff-Wuff brands. Both brands are mixtures of lamb, fish and beef compounds, the exact mix characterizing the differences between the two brands. The chart in Table 2 shows the amount of each of the compounds in a single package of the Bow-Wow and Wuff-Wuff brands, as well as the total amount of each compound that the company has available.

Compounds	**Totalavailable**	**Bow - Wow**	**Wuff - Wuff**
lamb	1,400 lbs	4 lbs	4 lbs
fish	1,800 lbs	6 lbs	3 lbs
beef	1,800 lbs	2 lbs	6 lbs

Table 2. *Ingredients needed for a package of Chow-Down dog food brands.*

Assume the Chow-Down company makes a profit of $12 on each package of Bow-Wow and $8 per package of Wuff-Wuff. Then the problem is how to use the available amounts of the three compounds so as to maximize their total profit.

To formulate this problem mathematically, let x_1 be the number of packages of Bow-Wow produced, while letting x_2 be the production of Wuff-Wuff. Since there are only 1,400 pounds of the lamb compound available, we must have

$$4x_1 + 4x_2 \leq 1400, \tag{1}$$

reflecting the first constraint in Table 1. Similar arguments based on the total amount of the fish and beef compounds available lead to the constraints

$$6x_1 + 3x_2 \leq 1800, \tag{2}$$

$$2x_1 + 6x_2 \leq 1800. \tag{3}$$

Since Chow-Down cannot produce a negative amount of either brand, we must also have $x_1, x_2 \leq 0$.

Note that these are all *linear* constraints, since the unknowns x_1 and x_2 appear to the first power everywhere. To complete the problem formulation, the profit P to Chow-Down can be expressed by the linear relation

$$P = 12x_1 + 8x_2 \tag{†}$$

The mathematical problem then is to find values for x_1 and x_2 that maximize P, subject to the above constraints.

The simplest way to see how to solve this problem is to draw a picture. Any pair of values for x_1 and x_2 constitute a point in the plane whose coordinates are (x_1, x_2). And since both quantities must be nonnegative, we can confine our attention to points in the first quadrant. Moreover, the three constraints (1)–(3) are all linear, so they can each be represented geometrically as straight lines in the plane. The overall situation is shown in Figure 2, where the constraints (1), (2) and (3) are shown as the lines ①, ② and ③ in the figure. The *feasible set*, consisting of all those points (x_1, x_2) in the first quadrant satisfying the constraints, is the shaded region OABCD, while the *boundary* of the feasible set is formed by those points on the hatched portions of the constraint lines. Finally, let's note the *vertex points*, which are the finite set of points where the line segments forming the boundary intersect.

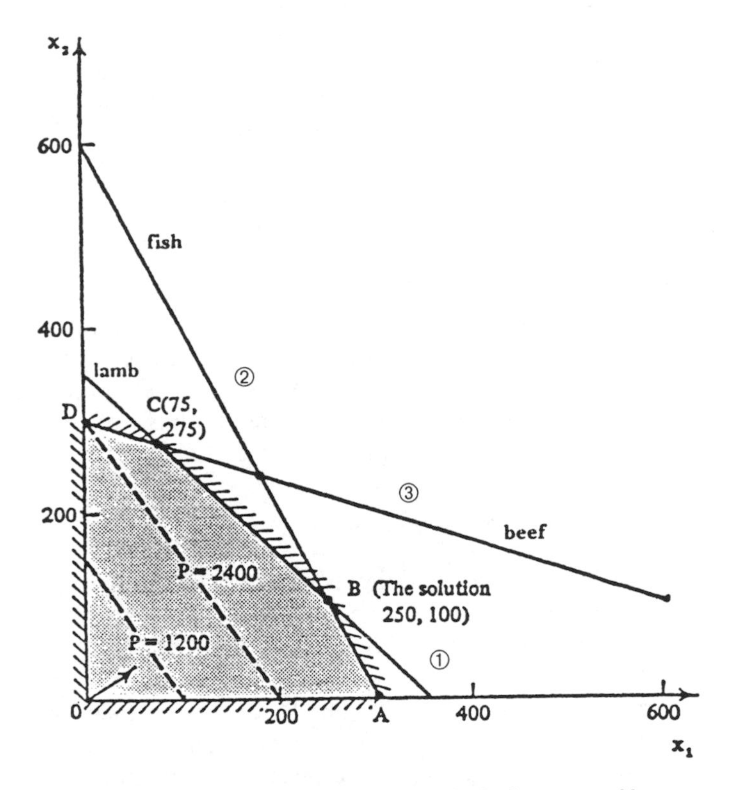

Figure 2. *Graphical solution of the dogfood mixing problem.*

The dashed lines in Figure 2 represent various constant levels of profits. So, for instance, all the points (x_1, x_2) on the dashed line labeled "$P = 1200$" represent production levels of the two brands of dogfood yielding a profit of \$1,200 for Chow-Down. The geometry of the feasible set, together with the fact that these level curves of constant profit are all straight lines, shows that the maximal profit must lie at one of the vertex points of the feasible set, in this case one of the points O, A, B, C or D. So by moving the constant profit line out from the point O as indicated by the arrow in Figure 2, the vertex point that is farthest from the origin and meets this profit line is the solution of the problem, i.e., it is the point of maximal profit satisfying all the constraints. As indicated in the figure, this turns out to be point B, which has the coordinates $x_1 = 250$, $x_2 = 100$. The profit at this point is $3,800. So if Chow-Down wants to get as much benefit as possible out of their supply of lamb, fish and beef compounds, they should produce 250 packages of the Bow-Wow brand and 100 packages of Wuff-Wuff.

This example illustrates the most important feature characterizing any LP problem: The solution is always found at a vertex point on the boundary of the feasible set. So, despite the fact that every point of the feasible set is, in principle, a candidate for the optimal solution, we really need only examine the vertex points to find the optimizer. The computational implications of this result are enormous, since the set over which we have to search for the solution is reduced from an infinite set (the feasible set of feasible points) to a finite set (the vertex points). Of course, this may still be a difficult computational problem. After all, the number of possible chess positions is also finite, but far beyond the capacity of any real or imagined computer to ever examine in a time less than the lifetime of the universe. Fortunately, however, the LP-case is not quite this bad. And there exist computational procedures for searching the vertex set in a computationally efficient and practically computable manner. Let's now look at the first of these procedures, the so-called *Simplex Method*, which still today forms the basis of most algorithms used in practice to solve linear programming problems.

The Simplex Method

In 1947, George B. Dantzig was working as a civilian in the Pentagon as a mathematical adviser to the Air Force Comptroller. As part of his job, Dantzig was often called upon by the Air Force to solve real planning problems, involving the way to distribute Air Force personnel, money, planes and other resources in a cost-effective fashion. Since most of these problems involved economics in one way or another, Dantzig enlisted the advice of economist Tjalling Koopmans about these linear programming problems, erroneously assuming that economists had developed solution techniques for them years before. To Dantzig's great surprise, Koopmans told him that the economists didn't have any procedures for systematically finding the solution to LP problems either. So Dantzig set out in the summer of 1947 to find one.

The first, and most important, step in Dantzig's search for a method to solve LP problems was the observation we noted in the dogfood-mixing problem. Namely, that the feasible region is a convex body — a polyhedral set like that shown in Figure 3. Therefore, Dantzig argued, the optimal point had to be at one of the corners of this set. Moreover, it would be possible to find improved values of the criterion function by moving from one such corner point to the next, much like the beetle in the figure as it crawls along the edges seeking, say, the point containing the greatest amount of food, marked here by the chocolate cake. In the jargon of algebraic topology such a polyhedral set is termed a "simplex", which gives rise to the name of Dantzig's algorithm for how the beetle should move along the edges in the most effective way to reach its goal.

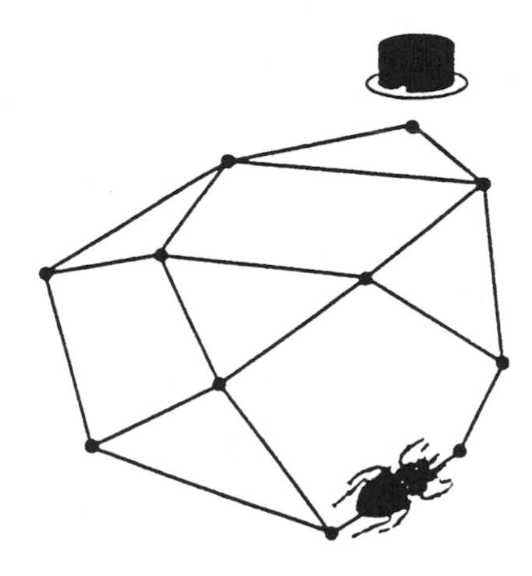

Figure 3. *A beetle crawling along the edges of a polyhedron.*

Originally, Dantzig thought such a procedure might be hopelessly inefficient, wandering along improving edges for a long time before reaching the optimal corner point of the simplex. But he was wrong! In fact, his empirical results indicated that it was difficult to find a problem involving m constraints in any number of unknown variables that could not be solved in m moves along edges. Thus was born

THE SIMPLEX METHOD. *Every linear programming problem can be solved by the following procedure:*

- *I. Find a vertex representing a feasible solution (i.e., one whose coordinates satisfy all the constraints of the problem), and calculate the value of the criterion function at that point.*
- *II. Examine each boundary edge of the feasible set passing through this vertex to see whether movement along such an edge will improve the value of the criterion function.*
- *III. If it does, move along the chosen edge to the new vertex point and evaluate the criterion function.*

IV. Repeat steps II and III until there is no longer an edge along which movement improves the criterion function. The current vertex is then the optimal solution to the problem.

So we see that the computational implementation of the Simplex Method involves two aspects: (i) finding a feasible solution to start the process, and then (ii) improving a feasible solution by moving along edges of the boundary of the feasible set from vertex-to-adjacent vertex, one step at a time, until reaching the optimal point.

With the development of linear programming, and especially the Simplex Method, we see a theory that is strongly oriented toward applied mathematics. In fact, both Koopmans and the Russian Leonid V. Kantorovich were jointly awarded the 1975 Nobel Prize in Economic Science for their work in exploiting and extending existing methods in the theory of linear inequalities versions of the general LP algorithm to solve very definite questions in resource allocation in large economic enterprises. Strangely, George Dantzig was *not* cited by the Swedish committee when it came time to hand out the Nobel accolades, despite his work laying the foundation for a general mathematical and computational procedure to solve all LP problems.

Lest the reader be too trusting of the distinction made in the last few sections between applied and motivated mathematics, relegating game theory to one camp and LP to the other, here is a short argument showing how the determination of optimal mixed strategies for a two-person, zero-sum game, strategies whose existence are guaranteed by Von Neumann's Minimax Theorem, can be formulated as a special type of LP problem.

Linear Programming and the Theory of Games

Suppose we have a two-person, zero-sum game of the type discussed earlier, having payoff matrix R. This means that if Player I chooses strategy i and Player II selects strategy j, the payoff to Player I is r_{ij}, while that to Player II is $-r_{ij}$. Now consider the problem faced by Player I, who must choose a probability vector p, whose ith component, p_i, is the probability of playing strategy i. With such a choice, Player I can be certain of receiving at least $\min_j \sum_i p_i r_{ij}$. Therefore,

$$r_{1j}p_1 + r_{2j}p_2 + \ldots + r_{mj}p_m \geq V, \qquad j = 1, 2, \ldots, n,$$

$$p_1 + p_2 + \ldots + p_m = 1, \qquad i = 1, 2, \ldots, m, \tag{¶}$$

Here we use the symbol V to represent the lower bound on what Player I can expect to receive from the game. Clearly, Player I wants to choose p so as to make V as large as possible. A similar argument from the perspective of Player II leads to a choice of the probability vector q so as to minimize the quantity $\sum_j r_{ij}q_j$, subject to the constraints $\sum_k q_k = 1$, $q_k \geq 0$, $k = 1, 2, \ldots, n$.

The Minimax Theorem establishes the existence of a number V satisfying these two optimization problems.

To reduce the game-theoretic calculation to an LP problem, we first add a constant to all the entries in the payoff matrix R to ensure that the value of the game V will be positive (for this, it suffices to add a number larger than $\max_i \min_j r_{ij}$). This increases the value of the game, but does not change the optimal strategies. Thus, we can safely assume that V is positive, and introduce the new variables $p_i' = p_i / V$, which will also then assume only nonnegative values. Dividing the inequalities of (¶) by V, we obtain the problem for Player I of minimizing over all $p_i' \geq 0$

$$\sum_{i=1}^{m} p_1',$$

subject to the constraints

$$\sum_{i=1}^{m} r_{ij} p_j' \geq 1, \quad j = 1,2,\ldots,n..$$

This is clearly a special LP problem in which the coefficients in the criterion all equal 1. Repeating this argument for Player II, we obtain the dual LP problem of maximizing over all $q_i' \geq 0$

$$\sum_{j=1}^{m} q_1',$$

subject to the constraints

$$\sum_{j=1}^{n} r_{ij} q_j' \leq 1, \quad i = 1,2,\ldots,m..$$

So what we end up with is a primal LP problem for the optimal strategy vector p for Player II, since the optimal value of the criterion is simply the sum of the p_i', which equals $1/V$. The dual problem is then used to determine the optimal strategy vector q for Player II.

At this point we could well ask, At what point does the *motivated* mathematics of game theory become the *applied* mathematics of linear programming? Or does it? Or is there even a difference? From the point of view of mathematics, there probably isn't. So let's return to the theme of creativity and see if there is any distinction to be drawn from that quarter.

CREATIVELY SPEAKING

The triadic view of creativity involves the interplay among the *person*, the *domain* and the *field* in assessing the creative content of a piece of work. Here we examine the above three pieces of mathematical work — general equilibrium theory, game theory and linear programming — from these perspectives.

The Person

Multiple intelligences and the individual's intimate relationships at the time of his or her work are the central foci of action at the personal level. In the work of Arrow and Debreu on equilibrium economics, it's easy to see both strong logico-mathematical abilities combining with the ability to see clearly the essentials of the world of economic affairs. Similar remarks apply to the pioneering results of Dantzig, Koopmans and Kantorovich on linear programming, as well as to the work by von Neumann and Morgenstern on game theory. In all the cases examined here, the individuals involved strongly confirm the thesis that creative work emerges out of an unusual combination of different intelligences, one for the mathematical formalism of the work, another for its relationship to the real world of economic affairs.

The triadic theory also suggests that most innovators are fiercely independent and basically loners, but that at the time of creative acts they have a strong need for a confidant. Unfortunately, the written record is rather skimpy on this aspect for the main actors in our story here — with the possible exception of von Neumann. Most accounts of von Neumann's life suggest that while he was quite gregarious on the surface, involving himself in many activities both inside and outside academia, his basic nature was quite reclusive. For example, in his biography of von Neumann [10], Norman Macrae notes that

> Even as a child, Johnny stood a bit outside ordinary social relationships . . . Johnny liked his classmates. He was anxious to feel easy with them but never achieved the ease he would have wished. Johnny was on the outside looking in, not in shyness or in envy, but he always felt himself more an observer than a participant.

And in his autobiography [11], Stan Ulam notes that these characteristics did not disappear in von Neumann's adulthood. For example, Ulam remarks that, "even at parties at his house, he [von Neumann] would occasionally leave the guests to go to his study for half an hour or so to record something that was on his mind".

Thus, insofar as these limited samples are concerned, we see the triadic theory of creativity supported in its claims creative genius tends to go hand-in-hand with the exercise of multiple intelligences wrapped up inside a loner's personality. Now what about the external factors, the area of application of this genius and the social dimension of the outside world?

The Domain

Referring back to Figure 1, we find that consideration of creativity and the domain centers on two principal aspects: the degree of structuring in the domain, and whether there exists a single dominant paradigm within the domain as opposed to two or more competing paradigms.

In regard to general equilibrium theory, the record is rather clear on both these counts. As with most areas of academic concern, economics is rather highly structured, with many levels between amateur fumblings and expert knowledge. Moreover, these levels are quite well delineated and agreed upon by experts. Equilibrium economics is no exception, either now or at the time Arrow and Debreu produced their pathbreaking work.

But the story is quite different when it comes to competing paradigms. According to the extensive study of Ingrao and Israel given in [12], in the late 1940s and early 1950s there were at least four recognizably distinct lines of research in equilibrium economics: (1) von Neumann and Morgenstern's game-theoretic approach, (2) a line of investigation due to Abraham Wald that emphasized the imposition of restrictions on the market's excess demand function, (3) Debreu's program for the axiomatization of economics, and (4) Paul Samuelson's focus on the goal of constructing economic theory on models of theoretical physics.

At just the opposite extreme was the situation in linear programming and economics, where the now well-recognized domain of "operations research" was only starting to emerge from work done during the Second World War. In the late 1940s, this work was an amalgam of economics, mathematics, and computing existing in a state very far from well-structured by anyone's definition. Moreover, there were several paradigms afloat at the time— linear programming, network analysis, dynamic programming, statistical decision theory, to name but a few. But it certainly cannot be argued that these paradigms were in any meaningful way in competition. While an area like dynamic programming might have more generality and be able to encompass a field like linear programming as a special case, the computational burdens associated with implementing the theory caused the whole domain of operations research to split into distinct camps concerned with particular classes of problems that could be solved by methods like the Simplex Method, tailor-made for the problems of that area.

Game theory is a singular situation. Here it's basically meaningless to even speak about the domain at the time von Neumann did his work on the Minimax Theorem, for the simple reason that there was no area called "game theory" until fifteen years after this work. The whole mathematical theory of games emerged almost full-grown with the magisterial work [13] by von Neumann and Morgenstern in 1944. So prior to this time, there was neither a domain, nor a paradigm.

The Field

The two areas of action at the field node are the size of the field and the degree of consensus that obtains within the domain. As far as general equilibrium theory is concerned, the field was fairly large at the time of the work by Arrow and Debreu, as reflected in the fact noted above that there were at least four quite distinct lines of research underway. Of course, the vast majority of economists at the time were occupying themselves with quite different matters pertaining to practical questions about restructuring the world's economy following World War II. But even in the rarefied heights of pure theory, it would not be fair to say that Arrow, Debreu et al were working in a vacuum. Similar remarks can be made about the work on linear programming, where many economists and mathematicians were at work on both the theory and application of "programming" problems.

Game theory, of course, is the odd man out. Here again von Neumann and Morgenstern were creating an entire field cut from whole cloth, and it was not until at least a decade later that disciples like Lloyd Shapley, Martin Shubik and John Harsanyi took up the challenge of developing the ideas in [13] into a full-fledged branch of modern mathematics.

On the matter of consensus, all three areas surveyed here — general equilibrium theory, game theory and linear programming — were recognized as major acts of creation especially by the economics community. We'll talk later about their reception by the world of mathematicians. And, as already noted, only game theory was not canonized by having its creator knighted with a Nobel prize. And this is surely only because von Neumann died before the Nobel Prize in Economic Sciences was added to the Nobel list in 1968 by the Central Bank of Sweden initiating the prize in economics as part of its 300th anniversary celebration. Thus, widespread consensus in the field of economics about the value of each of these theories is clear.

WHAT DOES IT ALL MEAN?

The American Mathematical Society Subject Classification Scheme is a way of categorizing papers in mathematics for summarizing in *Mathematical Reviews*. In this carving up of the mathematical forest into trees, papers are classified using a system that starts at 00 — xx (General) and goes to 98 —xx (Mathematical Education: Collegiate). Folk wisdom in the mathematical world has it that the status level of a topic is inversely proportional to the magnitude of the number the topic receives in this scheme. So, for example, papers on Logic and Foundations (category 02 — xx) rank much higher in the mathematical pecking order than those on Ordinary Differential Equations (category 34 — xx), which in turn greatly outrank material on Computer Science (category 68 — xx). So if one believes in the rank ordering of mathematical subjects by the AMS scheme, one way of assessing the creativity content of our three works of mathematical economics is to see where they sit in such a mathematical taxonomy.

The answer is quite clear: not very high. In fact, all three domains — general equilibrium theory, game theory and linear programming — are lumped together in the broad category 90 — XX (Economics, Operations Research, Programming, Games). More specifically, within this overall classification we find general equilibrium theory residing in subcategory 90A15 (Economic models), linear programming in 90C05 (linear programming) and game theory in 90DXX. Thus, on the basis of this fine distinction, the mathematical pecking order associated with our examples has general equilibrium theory just barely nosing out linear programming, with game theory bringing up the rear in what amounts to a three-horse photo finish.

So what can we conclude from this? Well, first of all, that the mathematical community doesn't attach much intrinsic mathematical merit to *any* of the three areas. And, secondly, what mathematical creativity does exist in the areas is almost indistinguishable, one domain from the other. Thus, on the basis of the "AMS test" for mathematical creativity, there is very little to choose among general equilibrium theory, linear programming and game theory. Hence, there is little to choose between applied or motivated mathematics either, at least to the degree that these examples typify both.

Taking an even more subjective, personal view of the question, I would argue that the greatest creative content in these examples lies with von Neumann's work in game theory and economics, primarily because it was created almost totally from scratch with few, if any, antecedents. Both general equilibrium theory and linear programming, on the other hand, were natural evolutionary outcomes of other work in economics, which gave people like Arrow, Debreu, Dantzig and Koopmans a foothold upon which to build their mathematical edifices. So, in this sense, the argument would fall in favor of the motivated mathematics of von Neumann as opposed to the applied mathematics of Arrow and Dantzig as having the greater creative content.

Surveying all the arguments given in this paper, it's difficult to escape the final conclusion that there is really no difference to be seen in the creative content of work in applied mathematics as opposed to motivated mathematics. In short, it is a pseudo distinction. Furthermore, there is also very little to discern in the creative process underlying these two approaches to the use of mathematics outside mathematics. Only the underlying forces giving rise to the original investigation differ. And that seems to be irrelevant to either the ultimate acceptance of the product by either the practitioners in the applied area or the mathematical community.

Santa Fe Institute, USA

REFERENCES

[1] Halmos, P., "Applied Mathematics is Bad Mathematics", in L. Steen, ed., *Mathematics Tomorrow*. New York: Springer, 1981, pp. 9–20.
[2] Von Neumann, J., "The Mathematician", in R. Heywood, ed., *The Works of the Mind*. Chicago: University of Chicago Press, 1947, pp. 180–196.
[3] Gardner, H., "Seven Creators of the Modern Era", in J. Brockman, ed., *Creativity*. New York: Touchstone Books, 1993, pp. 28–47.
[4] Cskiszentmihalyi, M., "Society, Culture, and Person: A Systems View of Creativity", in R. Sternberg, ed., *The Nature of Creativity*. New York: Cambridge University Press, 1988, pp. 325–339.
[5] Poincaré, H., "Mathematical Discovery", chapter 3, in Poincaré, H., *Science and Method*. Paris, 1908 (Dover reprint, n.d.g.).
[6] Hardy, G.H., *A Mathematician's Apology*. Cambridge: Cambridge University Press, 1940.
[7] Halmos, P., "Mathematics as a Creative Art". *American Scientist*, 56, 1968, 375–389.
[8] Arrow, K., "Social Choice and Justice", *Collected Papers*, vol. 2, Oxford: Basil Blackwell, 1984, p. vii.
[9] Debreu, G., *Theory of Value*. New Haven, CT: Yale University Press, 1959.
[10] Macrae, N., *John von Neumann*. New York: Pantheon, 1992, p.51, 73.
[11] Ulam, S., *Adventures of a Mathematician*. New York: Scribners, 1976, p. 78.
[12] Ingrao, B. and Israel, G., *The Invisible Hand*. Cambridge, MA: MIT Press, 1990.
[13] Von Neumann, J. and Morgenstern, O., *Theory of Games and Economic Behavior*. Princeton: Princeton University Press, 1944.

ÅKE E. ANDERSSON

CREATIVITY, COMPLEXITY AND QUALITATIVE ECONOMIC DEVELOPMENT

A GROWING INDUSTRIAL DILEMMA

The basic principles of work organization in the coming industrial society were clarified by Adam Smith in *The Wealth of Nations*, first published in 1776. And this book not only formulated the basic doctrine of the competitive economy and the constitutional rules needed for the smooth functioning of competition. To many the industrial revolution is perceived as a transformation from a society based on manual labor into a society with a production system based on machinery driven by manual based energy sources. However, this is a very minor aspect of Smith's vision. The core of his analysis is organizational. And the basic principle of industrial organization is *simplification by division of labor*. The creation of new ideas, products and industries is reserved for the entrepreneurial capitalists. The masses are relegated to operation of simplified processes, requiring a minimum of education. The rising productivity can be accomplished by a properly iterated use of people performing simple tasks:

To take an example, therefore, from a very trifling manufacture; but one in which the division of labour has been very often taken notice of, the trade of the pin-maker; a workman not educated to this business (which the division of labour has rendered a distinct trade), nor acquainted with the use of the machinery employed in it (to the invention of which the same division of labour has probably given occasion), could scarce, perhaps, with his utmost industry, make one pin in a day, and certainly could not make twenty. But in the way in which this business is now carried on, not only the whole work is a peculiar trade, but it is divided into a number of branches, of which the greater part are likewise peculiar trades. One man draws out the wire, another straights it, a third cuts it, a fourth points it, a fifth grinds it at the top for receiving the head; to make the head requires two or three distinct operations; to put it on, is a peculiar business, to whiten the pins is another; it is even a trade by itself to put them into the paper; and the important business of making a pin is, in this manner, divided into about eighteen distinct operations, which, in some manufactories, are all performed by distinct hands, though in others the same man will sometimes perform two or three of them. I have seen a small manufactory of this kind where ten men only were employed, and where some of them consequently performed two or three distinct operations. But though they were very poor, and therefore but indifferently accommodated with the necessary machinery, they could, when they exerted themselves, make among them about twelve pounds of pins in a day. There are in a pound upwards of four thousand pins of a middling size. Those ten persons, therefore, could make among them upwards of forty-eight thousand pins in a day. Each person, therefore, making a tenth part of forty-eight thousand pins, might be considered

Å. E. Andersson and N.-E. Sahlin (eds.), The Complexity of Creativity, 139–151.

as making four thousand eight hundred pins in a day. But if they had all wrought separately and independently, and without any of them having been educated to this peculiar business, they certainly could not each of them have made twenty, perhaps not one pin in a day; that is, certainly, not the two hundred and fortieth, perhaps not the four thousand eight hundredth part of what they are at present capable of performing, in consequence of a proper division and combination of their different operations. Adam Smith, (1776) *An inquiry into the nature and causes of the wealth of nations*, vol I.

Adam Smith's views on the higher education, scientific research and creative arts were restrictive and based on principles of calculation of human capital profitability, later on to be spelled out in detail by e.g. Gary Becker (1975).

Half a century after *The Wealth of Nations* the views of Adam Smith were complemented by an increasing number of economists writing on the advantages and disadvantages of the use of machinery and other equipment as a substitute for labor in tasks that were too heavy or strenuous for the human body. The role of creativity was at this stage coming a little more into the focus of economic analysts, as formulated by John Stuart Mill:

In a national, or universal point of view, the labour of the savant, or speculative thinker, is as much a part of production in the very narrowest sense, as that of the inventor of a practical art; many such inventions having been the direct consequences of theoretic discoveries, and every extension of knowledge of the powers of nature being fruitful of applications to the purposes of outward life. The electro-magnetic telegraph was the wonderful and most unexpected consequence of the experiments of Ørsted and the mathematical investigations of Ampère: and the modern art of navigation is an unforeseen emanation from the purely speculative and apparently merely curious inquiry, by the mathematicians of Alexandria, into the properties of three curves formed by the intersection of a plane surface and a cone. No limit can be set to the importance, even in a purely productive and material point of view, of mere thought. Inasmuch, however, as these material fruits, though the result, are seldom the direct purpose of the pursuits of savants, nor is their remuneration in general derived from the increase production which may be caused incidentally, and mostly after a long interval, by their discoveries; this ultimate influence does not, for most of the purposes of political economy, require to be taken into consideration; ... But when (...) we shift our point of view, and consider not individual acts, and the motives by which they are determined, but national and universal results, intellectual speculation must be looked upon as a most influential part of the productive labour of society, and the portion of its resources employed in carrying on and in remunerating such labour as a highly productive part of its expenditure. John Stuart Mill (1909) *Principles of Political Economy*, pp 41–42, new edition.

The consequences for society of the combined application of trade dependency, division of labor and newly invented machinery were to become spectacular. New cities were created and old manufacturing cities grew at sometimes staggering rates in terms of housing, population and employment. Meanwhile, new canals and railroads penetrated the countryside in the hunt for natural resources to be used as energy or raw material for the expanding manufacturing industries.

The industrial revolution developed in overlapping waves. The first wave of the industrialization process included Great Britain, Belgium and New England. The second wave moved industrialization to Germany, France and the Great Lake region of the United States. The third wave included less accessible parts of Europe, the United States and Japan by the second half of the 19th century. Currently the countries of South East Asia have become part of the fourth wave of the industrial revolution.

At each consecutive wave higher long-term growth rates of production have been recorded, because of the increasingly advantageous possibilities of imitation of production processes developed at earlier stages. History demonstrates a growth advantage of being backward.

There is an emerging dilemma of the progressively faster industrialization of the world. The free availability of natural resources has been a hidden assumption in the theory and practice of the growing industrial system. This assumption is becoming increasingly unreasonable. Even if resource depletion is not imminent, it has become obvious that resource conservation is necessary and would lead to increasing scarcity prices of most natural resources. Energy, whether generated by minerals, hydro power or other renewable resources, is also becoming increasingly scarce. Quantitative growth of industrial production, based on increasingly scarce natural resources obviously cannot go on forever. Is there a substitute for natural resources efficient enough for a sustained, although different character of, development? I think there is.

QUALITY AND CREATIVITY

The availability and accessibility of economic, technical and social data and other kinds of information stored in computer data banks and in printed form is expanding rapidly and seemingly without any upper limits to the growth. Accessibility of information and knowledge is improving by the decreasing prices of personal transportation and communication. Parallel to this process of expanding supply of information, there is since long a slow but steady expansion of knowledge, accumulated by research, education and on-the-job training. It should be stressed that information and knowledge are qualitatively different from each other. Although knowledge can be transmitted by books and other means of storage and communication it can hardly become a creative resource, unless it is embodied in some complete cognitive system.

Especially in the highly developed market economies there has been a steady decrease in the price of information and knowledge relative to the prices of energy, materials and uneducated labor. This calls for a reconsideration of the principles of production based on simplification by division of labor and the exploitation of natural resources, energy and unskilled labor. More specifically it requires the abandonment of quantitative growth in favor of qualitative development.

At this stage of the argument it ought to be remarked that any productivity measure is based on two elements. E.g. GNP (Gross National Product) is essentially proportional to the weighted sum of quantities of different goods and services, where the weights used are the prices of the different goods and services. Prices are thus defined as valuation coefficients. If these prices are properly measured so as to reflect valuations by the users, then obviously a growth of the GNP can be realized by growth of quantities *or by increasing prices of the products.* Prices, properly reflecting the quality of the products, are called *hedonic prices.* Thus, quantitative growth can be separated from qualitative development, as measured by the growth of hedonic prices at constant levels of production.

Hedonic prices are influenced by user-relevant characteristics of the product to be purchased. The hedonic price of a micro computer might for instance be a function of the ease of accessing programs, the speed of execution of algorithms, the readability and the elegance of display, the versatility of applications and the weight of the computer. These characteristics are obviously not identical with the characteristics relevant to the engineers and designers creating the new product. Creators of new qualities are primarily operating on a deeper level—i.e. on the blueprint or algorithm level. At that level the *complexity* of the product is determined. And it is the complexity of the product that will finally determine the characteristics relevant to the user of the product. Thus, indirectly the complexity of the product will determine the hedonic price to be offered.

Creativity can be more or less fundamental. A team of engineers and designers can be satisfied with the creative activity of changing the quantitative level of different characteristics within a *given characteristics space.* This level of ambition corresponds to what Margaret Boden has called *variational creativity* (1990). More fundamental creativity requires changes in the *dimensionality of the characteristics space.* The creation of the first bus by variations of characteristics inherent in the first car is thus an example of variational creativity within a given characteristics space. The only basic difference was the number of passengers to be carried. The creation of the first aeroplane was much more fundamental. It required at least one new dimension in characteristics space, added to the characteristics of the car, the bus and the lorry.

COMPLEXITY AND CREATIVITY

Any product, be it a musical score, a computer program, a car, or a micro computer, can be seen as a set of elementary entities or components and their interrelations. At some level of resolution any product can thus be seen as a more or less complex static or dynamic network of nodes and links connecting nodes with each other. Often nodes are within themselves networks. When a transformer is put into a computer as one of the nodes of the internal computer network it is in itself a network, created by some electrical engineers at some earlier stage of the creative process.

Complexity of a product is at least three dimensional. *The first and most obvious type* is algorithmic complexity. By this I mean the length of the shortest program or recipe that would solve a given problem or produce a copy of a given product.

Let us assume that the given "product" is some sequence of numbers on a paper e.g.:

1; 2; 4; 8; 16; 32; 64; ...

Then obviously the algorithm needed to produce a copy of this product would be the following:

$$x(r+1) = 2 \cdot x(r);\ x(0) = 1$$

This is a very short computer algorithm. Objectively, the series of numbers given is of rather low complexity, according to our definition. Subjectively, the series itself is also not perceived as very complex.

It does not take a lot of creativity to discover the inherent algorithmic structure of the consecutive numbers. To go from the *ex ante* set of numbers to the discovery of the mechanism or algorithm, that would exactly produce a copy of this string of numbers is not hard. The *ex ante* complexity of the problem is very close to the *ex post* complexity of the computer algorithm. The difference between *ex ante* and *ex post* complexity could qualify as one measure of the creativity involved in discovering the solution.

Another example would clarify this point. Let us assume that we come across the following series of numbers:

1; 2; 3; 5; 2; 4; 4; 3; 5; 1; ...

Obviously, most of us would agree that this series of numbers have a high degree of *ex ante* complexity. It is very hard to figure out what mechanism or algorithm that would have generated this series of random-looking numbers. To find the deterministic algorithm that would produce an exact copy is either very tedious or would require some creativity, if we want to arrive at the shortest possible algorithm (mechanism or machinery) that would be capable of producing exact copies of the series. There exist such an algorithm, namely the tent map on p. 140.

This tent map consists of one expansion phase up to a certain critical point x_c. After that point the process is contractive. Writing it out as an algorithm would require approximately twice as long a program as the one needed to reproduce the first series of numbers. Although this string of numbers is perceived as much more complex than the first string of numbers, the resulting algorithm is of surprisingly small complexity. But the difference between *ex post* and *ex ante* complexity is very large.

An important part of creativity lies in the capacity to bridge the gap

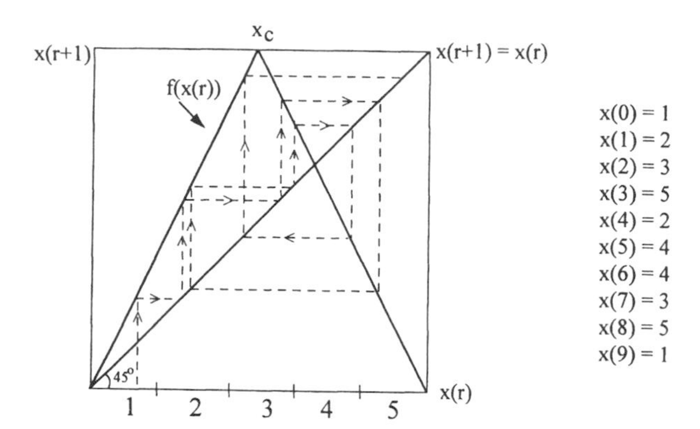

between *ex post* and *ex ante* complexity.[1] To detect that seemingly random numbers are not random at all but generated by some reliable, possibly deterministic, mechanism is one of the signs of creative research. Kepler's equations of planetary motions and Newton's gravity model are examples of how creativity of the scientists was capable of reducing the difference between *ex post* and *ex ante* complexity.

But this is also true of the more mundane creation of new products. When the brothers Wright created the first flying vehicle, heavier than air, the core of their invention was the discovery that controllability requires structural instability. Only by *not* mimicking the stability properties of flying birds was it possible for them to arrive at a mechanism that would require very little of human energy to be controllable.

The second type of complexity is the dimensionality of the input vector. E.g. in the preparation of a standard pea soup, according to some recipe, there is a need for some inputs. The number of different inputs is, however, rather limited. If we would instead aspire to produce a kettle of real first rate bouillabaisse, we would not only be forced to struggle with a much more complex recipe. We would also have the problem of importing and in other ways acquiring the large number of different inputs needed for the production of the ideal bouillabaisse.

The third type of complexity is associated with the human knowledge or skill needed for the actual use of the algorithm (recipe) and the different inputs in the actual generation of the required product. Some of the skill is reducible into algorithmic knowledge and can thus be transformed into some software (possibly of high algorithmic complexity) to be reembodied in a computerized machine. Mostly, part of the skill is tacit knowledge, irreduc-

[1] This amounts to the discovery of the underlying strange attractor which ordinarily requires a large number of consecutive observations.

ible from the human embodiment. The skills of Stradivarius in producing violins and cellos were most certainly of the latter kind. The complexity of knowledge embodied in humans is an important dimension of complexity of products. The minimum time of supervised apprenticeship and on-the-job training could be a possible measure of this third type of complexity of a product, generated in a specific production process. The irreducible and non-transferable character of this type of knowledge was probably what the world champion of skiing, Ingemar Stenmark, referred to when he was asked about the secrets behind his successful downhill skiing. His answer was "It is no point in explaining to someone who doesn't know".

THE CREATION OF ECONOMICALLY OPTIMAL COMPLEXITY

In the search for higher user value of products there seems to be a necessity of moving in the direction of increasing complexity, if an increased willingness to pay is to be accompanied by decreasing inputs of energy and raw materials. The development of computer technology is a salient example of this process. The earliest computers were extremely bulky and the internal complexity of the product was rather limited in all the three dimensions, defined above. It was in fact even possible for the user of the first computers to repair the computer, if it would break down. The newest versions of PCs and Power Macs are far too complex to admit users as repairmen.

There are of course limits to an economically motivated search for increased complexity. Increasing the levels of complexity of a given product requires the use of limited research and development resources. Industry tends to regard costs of creativity as reasonable, only if they would be followed by a larger increase in the revenues from sales of the improved product.

Disregarding the problems of uncertainty of research and development efforts we could formulate the optimality problem as in the following stylized decision model of a firm.

Maximize $G = p(z(c)) \cdot q(v) - T(c,r) - \omega \cdot v$

z = vector of product characteristics

$c = \{c_1, c_2, c_3\}$ = vector of complexity variables

p = hedonic price of the product, assumed to be concave in z and c

q = volume of production

v = vector of input quantities

$T(c, r) = 0$ is a convex transformation function regulating the possibilities of transforming research efforts (r) into complexities (c).

The necessary conditions of optimality require:

$$\sum_j \frac{\delta p}{\delta z_j} \cdot \frac{\delta z_j}{\delta c_i} \cdot q^* - \frac{\delta T}{\delta c_i} = 0; \; (i = 1,2,3)$$

$$p \cdot \frac{\delta q}{\delta v_s} - \omega_s = 0; \; (s = 1,\ldots,n)$$

The first optimality condition can be rewritten as:

$(a) \sum_j \frac{\delta p}{\delta z_j} \cdot \frac{\delta z_j}{\delta c_i} = -\frac{\delta T/\delta c_i}{q^*}$; or The marginal hedonic price increase should equal marginal opportunity cost of complexity per unit of production
where q^* = the optimal scale of production
and = the marginal opportunity cost of improving complexity in the i-th dimension.

Condition (*a*) shows that there are economies of scale in the increase of complexity of a product. This would be a factor explaining the industrial concentration of research and development efforts within the larger corporations, primarily.

CONNECTING NETWORKS OF IDEAS

Nothing is created in a social or economic vacuum. New ideas tend to be offsprings, variations or combinations of earlier ideas. Hofstadter (*Gödel, Escher, Bach —An Eternal Golden Braid*, 1979) has made the variation on themes the most essential aspect of creativity. This combinatorial aspect of the creative process can be put into a network context.

We can use the potential of this perspective within the context of modelling the development of scientific or artistic ideas. Let us assume that such ideas are accumulated in nodes of some physical or intellectual network. Let us further assume that some of these ideas are of a paradigmatic nature, thus by definition changing extremely rarely.

We can see the Newtonian paradigm of the sciences or the modernist paradigm of the arts as slow change processes within such networks of ideas. Presuming the components of such paradigmatic, fundamental ideas to be interconnected with each other in a stable network of consistent ideas, there can indeed be a very fast process of variational generation of more or less tangible models and similar artifacts borne by the "fundamental theory network". We can illustrate the idea with the following diagram:

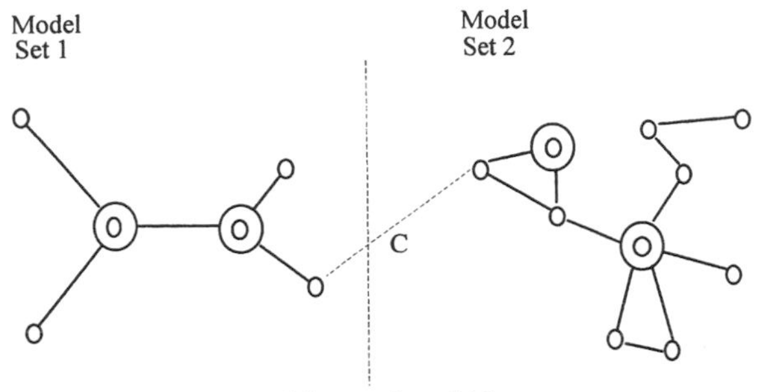

Network of ideas

In a way which is similar to the creation of a physical network, there can be critical as well as non-critical ideas generated. A critical idea must by necessity belong to the slowly changing part of a theory (or an artistic paradigm). A critical idea is one which interconnects large subsets of the total network of ideas, thus generating a larger foundation (or potential) for further theoretical or artistic development.

Such a critical idea (lemma, model, hypothesis, sketch, etc.) has two characteristics. Firstly, it provides a shortcut between formerly unconnected networks of ideas. Secondly, despite being small in itself, it triggers a significant increase in the development potential of the field as a whole. Both of these phenomena introduce fundamental non-linearities into the theories of art and science. This is so, because the introduction of a small, but sufficiently critical idea (a small increase in one of the slowly changing variables) leads to a large shift in the whole structure of ideas. This means that marginal evaluation is not meaningful in the vicinity of a critical link.

There are many examples of critical, but seemingly insignificant, contributions which have triggered the development of totally new approaches to model building or artistic work. One example is the book *Inequalities* by Hardy *et al.* (1959) in which the three mathematicians put forward the idea of reformulating many relations as inequalities rather than as equalities (which had been the standard procedure until then). In a sense, this approach was the critical link necessary to open up the use of set theory in practical problem formulation and problem solving. Modern mathematical programming needed this critical link between applied mathematical modelling and pure mathematical theory.

In the same way, Alberti's book *Della Pittura* (1435–36) had a similar triggering effect in connecting classical geometry with the creative efforts of the pictorial arts. When read in modern times, the critical idea seems simply to involve the introduction of an obscurely demonstrated, but basically correct, idea of perceptionally reasonable projection from three to two dimensions. Through this simple device, however, the whole world of geometry was brought into contact with the pictorial arts. Small causes with enormous consequences!

CREATIVITY VERSUS PRODUCTIVITY

Simplification by division of labor is a central productivity enhancing mechanism of the industrial system. The conflict between creativity and productivity by increasing simplification of production processes were discussed already by Adam Smith in volume II of *The Wealth of Nations*:

In the progress of the division of labour, the employment of the far greater part of those who live by labour, that is, of the great body of the people, comes to be confined to a few very simple operations; frequently to one or two. But the understandings of the greater part of men are necessarily formed by their ordinary employments. The man whose whole life is spent in performing a few simple operations, of which the effects too are, perhaps, always the same, or very nearly the same, has no occasion to exert his understanding, or to exercise his invention in finding out expedients for removing difficulties which never occur. He naturally loses, therefore, the habit of such exertion, and generally becomes as stupid and ignorant as it is possible for a human creature to become. The torpor of his mind renders him, not only incapable of relishing or bearing a part in any rational conversation, but of conceiving any generous, noble, or tender sentiment, and consequently of forming any just judgment concerning many even of the ordinary duties of private life. Of the great and extensive interests of his country he is altogether incapable of judging; and unless very particular pains have been taken to render him otherwise, he is equally incapable of defending his country in war. The uniformity of his stationary life naturally corrupts the courage of his mind, and makes him regard with abhorrence the irregular, uncertain, and adventurous life of a soldier. It corrupts even the activity of his body, and renders him incapable of exerting his strength with vigour and perseverance, in any other employment than that to which he has been bred. His dexterity at his own particular trade seems, in this manner, to be acquired at the expense of his intellectual, social, and martial virtues.

But in every improved and civilized society this is the state into which the labouring poor, that is the great body of the people, must necessarily fall, unless government takes some pains to prevent it.

It is otherwise in the barbarous societies, as they are commonly called, of hunters, of shepherds, and even of husbandmen in that rude state of husbandry which precedes the improvement of manufactures, and the extension of foreign commerce. In such societies the varied occupations of every man oblige every man to exert his capacity, and to invent expedients for removing difficulties which are continually occurring. Invention is kept alive, and the mind is not suffered to fall into that drowsy stupidity, which, in a civilized society, seems to benumb the understanding of almost all the inferior ranks of people. Adam Smith, (1776) *The Wealth of Nations* vol II.

The focus on productivity has a price in terms of losses of creativity. In the transformation from an industrial society into a society based on increasing value and complexity of products, human creativity will be increasingly important. Meanwhile, robotization and computerization of production will limit future advantages of specialization by division of labor. Synergy will replace division of labor.

CREATIVE ORGANIZATIONS

Most studies of productive industrial organization tend to superimpose a hierarchy upon an assignment of specialized and simplified jobs according to each individuals comparative advantage. If the productivity of each worker is independent of the pattern of assignment of other workers, i.e. if productivity is separable, then all workers should be assigned to the task within which he is comparatively most productive. It can be shown that there exists a set of wage rates for each type of specialist worker in such an organization, that would exactly support the most productive assignment of workers to different tasks.

It has also been shown that the optimal organization of control in such a productively organized industrial firm must necessarily be a hierarchy (Beckmann 1978).

Both the principle of assignment to tasks according to comparative advantages and the organization of control by hierarchy are severely limited within an organization, dependent upon maximal creativity of the employees. The principles of creativity by synergy of ideas, embodied in different persons, effectively limits the degree of separability of workers. If separability can be achieved, it will only be possible at the level of a group or cluster of individuals.

A study by Pravin Varaiya (1989) of the process of creating the IBM product ProPrinter has shown that almost continuous self-reorganization was needed for the successful completion of this creative project. The project also showed that the focus of control was shifting between different persons during the creative process of a group. No permanent hierarchical system of control was ever found.

SOCIAL CONDITIONS OF CREATIVITY

Very few systematic studies of macro-social conditions of creativity do exist. However, in Andersson (1985) a number of historically creative regions were studied from this point of view. Some examples of such regions are Miletos, Alexandria and Athens of the classical period, Florence and Bruges of the Renaissance, Amsterdam and London of the 17th century and Vienna, Berlin and Stockholm in their early period of industrialization. Among contemporary regions of creativity Los Angeles and San Francisco could be mentioned as important examples. There are some communalties of these examples of creative societies.

Firstly, creativity cannot be organized so as to be of the same duration as productivity. Creativity tends to be a socially short-lived phenomenon. All of these creative regions seem to have developed on the borderline to a state of severe structural instability.

Secondly, creativity tends to go hand in hand with expansion of communication with other regions. McClelland (1958) has shown that the creative expansion and contraction of Athens coincided with the expansion and contraction of the Athenian network of trade in the Mediterranean. The expansion of creativity in Florence and Bruges came alongside a peek in the influence of traders and bankers of these regions in Europe. And these communalities have tended to be reinforced with the expansion of the world economy. Closed societies are not conducive to creativity.

Thirdly, creativity seems to breed on diversity of knowledge, experiences and culture. All of the regions mentioned above have been at the crossroads of different cultural, ethnic and intellectual groups.

Fourthly, extreme inequality of wealth seems to be conducive to the financing of often expensive and uncertain creative experiments. This factor would operate in favour of creativity in two different ways. The first way is simple and obvious. Basic creativity in the arts and sciences is time consuming and tends to require an agglomeration of a mass of artists or scholars before the (uncertain) creative expansion occurs. The second reason for the need for inequality of wealth is associated with the social dilemma of the *nouveaux riche*. Creators of wealth are rarely socially accepted. General support to creative scientists or artists is often an efficient means of acquiring social recognition. Obvious examples are Lorenzo di Medici, Alfred Nobel and François Mitterrand.

CONCLUSIONS

The highly developed economies of the world are entering a new stage of development after centuries of unprecedented growth of productivity. The growth of productivity was generated by a combination of increasing division of labor, uninhibited use of energy and other natural resources and an expansion of the use of mechanical equipment in the production of essentially simple commodities. This expansion trajectory is now perceived as unsustainable. Further economic development has to be based on improvement of the quality of products accompanied by a much slower rate of quantitative growth.

Increasing consumer valuation of products at decreasing use of energy and other natural resources can only be achieved by increasing product complexity. But such growth of the inner complexity of products requires organization for creativity rather than for productivity. And increasing creativity is not compatible with increasing division of labor by simplification and specialization of labor. Quite to the contrary, creativity requires synergy of ideas, often embodied in different persons. Division of labor is thus inherently counter-creative.

The principle of division of labor has been carried beyond the individual work place in the development of the industrial society. Geographical zones, economic regions and even nations have become specialized on a few industrial and agricultural products so as to achieve the best preconditions for growth of productivity. Studies of creative regions have shown that such regions cannot be centers of creativity. Creative regions tend to be culturally, ethnologically and socially open societies. Creative regions are globally interactive regions.

Institute for Futures Studies, Sweden

REFERENCES

Alberti L.B. (1435–36) *Della Pittura* [English translation: (1966) *On Painting*, Yale University Press, New Haven, Connecticut].

Andersson Å. (1985) *Kreativitet – StorStadens Framtid*, Prisma.

Becker G.S. (1975) *Human Capital*, Chicago.

Beckmann M (1978) Rank in Organizations. Lecture Notes in Economics and Mathematical Systems, Springer-Verlag, Berlin-Heidelberg-New York.

Boden M. (1990) *The Creative Mind: Myths and Mechanisms*, Weidenfeld & Nicolson, London.

Hardy G., Littlewood J., Polya G. (1959) *Inequalities*, Cambridge University Press, Cambridge, MA.

Hofstadter D. (1979) *Gödel, Escher, Bach —An Eternal Golden Braid*, Basic Books Inc.

McClelland D.C. et al. (1958) *Talent and Society*, New York.

Mill J.S. (1909) *Principles of Political Economy*, pp. 41–42, new edition.

Smith A. (1776) *An inquiry into the nature and causes of the wealth of nations.*

Varaiya P. (1989) “Productivity in Manufacturing and the Division of Mental Labor” in *Knowledge and Industrial Organization*, eds. Andersson, Batten & Karlsson, Springer-Verlag.

THE COMPLEXITY OF CREATIVITY

Edited by

ÅKE E. ANDERSSON

Institute for Futures Studies,
Stockholm, Sweden

and

NILS-ERIC SAHLIN

Department of Philosophy,
Gothenburg University
and
Lund University,
Sweden

This is a book on the concepts, theories, models and social consequences of creativity. It contains articles written by well-known scientists and philosophers. It is not primarily a textbook, but can preferably be used both for undergraduate as well as graduate seminars. What makes this volume special is that it brings together the views of scientists from rather different disciplines on a very important topic – CREATIVITY. As far as we know there is no competing volume, this is the only one of its kind (Boden's volumes are related).

Audience

General philosophers of science and scholars working in cognitive science and psychology. It would also be of interest to economists, philosophers, psychologists and people interested in creativity.